Palgrave Studies In Play, Performance, Learning, and Development

Series editor
Lois Holzman
East Side Institute for Group
and Short Term Psychotherapy
New York, NY, USA

This series showcases research, theory and practice linking play and performance to learning and development across the life span. Bringing the concerns of play theorists and performance practitioners together with those of educational and developmental psychologists and counsellors coincides with the increasing professional and public recognition that changing times require a reconceptualization of what it means to develop, to learn and to teach. In particular, outside of school and informal learning, the arts, and creativity are coming to be understood as essential in order to address school failure and isolation. Drawing upon existing expertise within and across disciplinary and geographical borders and theoretical perspectives, the series features collaborative projects and theoretical crossovers in the work of theatre artists, youth workers and scholars in educational, developmental, clinical and community psychology, social work and medicine—providing real world evidence of play and theatrical-type performance as powerful catalysts for social-emotional-cognitive growth and successful learning.

Advisory Board

More information about this series at
http://www.springer.com/series/14603

Jaime E. Martinez

The Search for Method in STEAM Education

palgrave
macmillan

Jaime E. Martinez
New York Institute of Technology
New York, NY, USA

Palgrave Studies In Play, Performance, Learning, and Development
ISBN 978-3-319-85757-2 ISBN 978-3-319-55822-6 (eBook)
DOI 10.1007/978-3-319-55822-6

Printed on acid-free paper

This Palgrave Macmillan imprint is published by Springer Nature
The registered company is Springer International Publishing AG
The registered company address is: Gewerbestrasse 11, 6330 Cham, Switzerland

Acknowledgements

I would like to express my thanks to all the people who provided the support and encouragement for my performance of writing a book.

I would like to thank my sister Christine Martinez for getting me through a major revision in a short amount of time.

I would like to thank Samantha Ostrow for transcribing all the interviews and Lindy Judge for reading and copyediting the final manuscript.

I would like to thank Lois Holzman for directing my performance of writing a book.

I would like to thank Amy Bravo, Ellen Darensbourg, and Lauren Rigney. Without each of you, we would never have created the service-learning project that I am sharing in this book.

I would like to thank my students, colleagues, and friends for contributing their voices and ideas about STEAM education.

Last and not least: I would like to thank my lovely wife Migdalia for all things that she did that made it possible for me write this book.

Contents

PART I

Setting the Stage

Introduction

This book is about creating new kinds of developmental interdisciplinary learning environments that will be necessary if STEAM (Science, Technology, Engineering, Art, Math) education reforms are going to amount to more than just another educational fad. The title is a "shout out" to Lev Vygotsky, a developmental psychologist whose ideas I will be referencing throughout the book. Vygotsky contributed important and transformative theories of learning to the education research field and used the phrase "search for method" to describe his efforts to discover a new psychology, one that would be helpful to people during a revolutionary moment in Soviet history in the 1920s and 1930s (Vygotsky 1978, p. 65).

At a time when educators, educational researchers, and policymakers are trying to figure out how to use traditional knowledge acquisition methods of education to create STEAM education, I am concerned with transforming learning environments into ones that are developmental and interdisciplinary. In writing this book, my approach has been more creative than academic, and the data I offer is in the dialogues and stories. The voices of educational innovators who are creating and collaborating beyond the disciplinary boundaries of the institutions they work for will be prominent. I find that conversations and stories are a great way to learn developmentally.

Throughout, I provide accounts of practitioners who are creating developmental STEAM and STEM learning environments in schools and colleges. The projects featured are not traditional research studies; they share changes to teaching practice brought about by the specific needs of educators and their students. This book is the result of my search to create a new way of envisioning how a transformation of educational institutions might be possible. I believe institutional change is possible, not as a legislative policy innovation, but as something that will happen organically if we build the educational alternatives that are needed.

Some Background

Since 2009, my life has changed in many ways. I've graduated from a doctoral program in Urban Education, published portions of my doctoral research in my first book, *A Performatory Approach to Teaching, Learning and Technology* (2011), and I also received my first appointment as a university professor. I am currently working in the School of Interdisciplinary Studies and Education at the New York Institute of Technology (NYIT). I teach teachers in a graduate-level program in instructional technology. I consider myself a Vygotskian practitioner/researcher, a community organizer, and a performance activist. According to Lois Holzman's Social Therapeutics blog, "performance activists are part of an emerging global movement for social transformation through the use of play and performance—both in the form of 'plays' and in how people live their day-to-day lives" (Holzman 2013).

Both my first book and this one are efforts to share my development as an educator using Vygotskian approaches to teaching. I am a New York State-certified teacher and have taught elementary school children in underserved communities in the South Bronx in New York City, middle school students in a high-performing school in Manhattan, and poor children as far away as rural Nicaragua.

My formal training in teaching began with the New York City Teaching Fellows program in 2002 and continued as a doctoral student at The Graduate Center at the City University of New York. The training was focused on teaching and learning in urban school settings impacted by poverty and which have highly diverse student populations from different cultures in densely populated areas. My training in Vygotskian performatory approaches to learning began in 1996, outside of academia, in a youth development afterschool program (Martinez, 2011). It was that training, and a variety of other developmental learning experiences, that helped me transform from an information technology professional with no interest in education into an educator. My research interests focus on Newman and Holzman's Vygotskian cultural performatory approaches to human development (Newman and Holzman

1997). My projects feature STEAM/STEM education, academic service-learning (an educational program that brings college students to communities to provide voluntary public service), performance and play in institutional settings, and teacher learning and development.

Newman and Holzman

Dr. Fred Newman is a Stanford-trained philosopher, and Dr. Lois Holzman is a developmental psychologist. They are the developers of social therapy, a group therapy that creates opportunities for social growth and emotional development (Holzman and Mendez 2003). Traditional therapy is focused on the growth of the individual and discovering root causes of problems. Social therapy is concerned with the development of the group. In social therapy, people learn that when they contribute to groups, they discover new ways of seeing and being, and they create their emotional development. Newman and Holzman have collaborated on books, academic papers, conference presentations, and community-based projects for over 40 years. Their work is creative and community-based, independently funded, and outside of academia. Expansions to the key methods of social therapy have emerged in the practice of creating afterschool programs, community mental health programs, avant-garde political theater, grassroots independent political movements, and K-12 community-based schools (Holzman 1997, 2009; Newman and Holzman 1997). Over the years, I've attended numerous training seminars, online courses, and workshops, and most recently, I participated in a year-long residency program on social therapeutics. I have been active in the community development projects that Newman and Holzman have created, and I've used social therapy in my own life for personal growth and development. Newman and Holzman's Vygotskian cultural performatory approach makes up the theoretical framework for the community service-learning projects that I am involved in and my approach to teaching and learning. In this book, I use performatory approaches for understanding the unique opportunities of the STEAM education movement and creating new kinds of responses to old problems in learning at school.

Vygotsky's Search for Method

Vygotsky's efforts in the area of human learning and development were cut off when he died at age 38. The Stalinist Soviet regime of the 1930s suppressed efforts by students and colleagues to continue his work.

Nonetheless, Vygotsky's writings were preserved, and decades later in the 1960s came the first English language translation. A short compilation of some of his writings titled *Mind in Society* (1978) launched subsequent decades of flourishing Vygotskian study that continues to this day. According to Lois Holzman, the original English translation in the 1960s did not cause a stir, but *Mind in Society* did (L. Holzman, personal communication, December 16, 2016). That book describes the methodology of Vygotsky's search, his efforts to understand how human beings learn, and how the inner workings of the mind are transformed in social interaction. He was searching for a new scientific, social psychology.

> The search for method becomes one of the most important problems of the entire enterprise of understanding the uniquely human forms of psychological activity. In this case, the method is simultaneously pre-requisite and product, the tool and the result of the study (Vygotsky 1978, p. 65).

Vygotsky (1978) determined that "process" was the proper unit of study for his investigations. He referred to his method as "experimental-developmental" (p. 61). He understood that the experiments he was conducting "provoke" or "create" psychological development (p. 61). Vygotsky's unique contribution was this new conception of experiment and development mutually defining each other. The types of psychological phenomena that Vygotsky was interested in required a different kind of tool. He tells us that the method and the object studied are together in a dialectical relationship, or "tool and result."

Newman and Holzman's search for method differs. In their words, "[u]ltimately, we complete Vygotsky by rejecting his project for creating a sociocultural, scientific psychology, abandoning psychology altogether for an unscientific cultural (Performatory) approach to a practical-critical understanding of human life." (Newman and Holzman 2006, p. 9). Newman and Holzman are creating a new "unscientific cultural (performatory) approach" to "practical-critical" understanding. And what I am undertaking in this book is a performatory approach to understanding STEAM education. Some readers may want to apply scientific understandings to the projects and ideas in this book; however, it will be important to keep in mind that my approach is performatory.

Zone of Proximal Development

Vygotsky challenged psychology by suggesting that learning happened among people engaged in social activities together in zones of proximal development (ZPD). "It is the distance between the actual developmental level as determined by independent problem solving and the level of potential development as determined through problem-solvingunder adult guidance or in collaboration with more capable peers"; (Vygotsky 1978, p. 86). Vygotsky describes a process for learning that includes people of different abilities and knowledge working together in a problem-solving activity. According to Holzman, "The process of learning and the product of learning are created together" (Holzman 2008, p. 8). She notes that the conception of the ZPD that she and Newman have developed "parts ways with the more typical interpretation that it is an interactionist scaffolding process that aids in enculturation" (p. 18). Holzman also describes the ZPD as "simultaneously the performance and the performance space" (as cited in Holzman 2008, p. 19). For me, one of the central challenges in this book was to imagine, direct, and produce performances of interdisciplinary STEAM development in traditional educational settings.

Performance and Play

> Play, understood as performance, is being who you are not. Performance, understood as development, is creating who you are by being who you are not. Development, understood as relational activity, involves the continuous creating of stages (ZPDs) on which one performs "oneself" through incorporating "the other." (Holzman 1997, p. 73)

I view everything that I do in my life as a performance. The performances that I, and others, create are always changing and developing. My projects with students and communities are designed to encourage "being who you are not" or creating a process of "being and becoming" (Holzman 2009, p. 17). Part of understanding the concept of a ZPD requires looking at "the role of imitation in learning" (Vygotsky 1978, p. 87). Holzman describes creative imitation as "relating to oneself as/being related to by others as/performing as a speaker, a dancer, a writer, a learner, a human being. It is how children are capable of doing so much in collective activity" (Holzman 2009, p. 30). When a child has opportunities to interact with other kinds of people, there are more opportunities for creative imitations of many kinds. Creative imitation is an important idea that plays a significant role in the design of service-learning projects in Part III of this book.

Play is important to understanding ZPDs, and Vygotsky believed that it was in play that a child performs "a head taller" than her actual level of development and that "play contains all developmental tendencies in a condensed form and is itself a major source of development" (Vygotsky 1978, p. 102). Holzman refers to Vygotsky's writings on play as the inspiration for "linking creative imitation with performance, and performance with the dialectic of being/becoming that is development" (Holzman 2009, p. 31). In her keynote address at the North Carolina Honors Association Conference, Holzman says, "As a play revolutionary, I believe that play can revolutionarily transform the world and all of its people" (Holzman 2013, p. 1). I believe this also, and I try to incorporate "playfulness" in my "work" to transform the work.

My experience in institutional settings such as schools is that the rules sometimes make it seem that attempting a change or innovation is impossible. People are understandably concerned when institutional rules are broken. I have found that it is possible to play within the rules. On occasion, it is even possible to play with the rules. The transformative projects featured in this book tend to do both, and most important, they create new situations for which there are no rules.

Scientific Paradigms

Reading and listening to the audio book version of *The Structure of Scientific Revolutions* by Thomas S. Kuhn, originally published in 1962, created a significant, "aha" moment in my efforts to understand how creating developmental learning environments connected to interdisciplinary education in STEAM disciplines. Kuhn describes what scientific paradigms are (the shared ways that scientists see) and how paradigms rise and fall (what scientists do about unresolved problems and anomalies within a paradigm). He provides the example of how Newtonian physics (a paradigm) eventually gave way to Einstein's theories of relativity (a different paradigm). Kuhn suggests that the idea of a paradigm is applicable outside of science and that our commonly accepted practices, our ways of seeing the world, can be described using the lens of various paradigms rising and falling and moments of revolutionary shifts. I believe that the field of education is at a moment when the current paradigm is shifting. In the USA, many of the practices and theories of education that we take for granted are oriented toward workforce development and national economic competitiveness.

If I apply the idea of a paradigm to education, then social and institutional arrangements become a little clearer. To improve the effectiveness of the current education paradigm, senior members of the community of educators and education policymakers reinforce the use of education research and standardized teaching practices. They encourage practitioners to solve well-defined problems in education in the interest of national priorities. Researchers and practitioners must follow the mandates of the institutions they work for and may be further incentivized to focus on national priorities and standardization with grant funding. Any changes in practice, such as those brought about by the solution of a problem in education, would be considered an advance, reform, or refinement of existing practices. The role of institutions such as schools of teacher education and state certification boards is to ensure that newcomers to the profession are properly prepared to take up standardized and research-based practices. Mainstream education research activity contributes to advancement, refinement, and reforms. Mainstream practices are not intended to lead new practitioners to break from well-established practices. Interestingly, according to Kuhn, it is early career practitioners and other outsiders who typically attempt to introduce the practices of a revolutionary paradigm shift.

Developing Development

In *The End of Knowing: A New Developmental Way of Learning* (1997), Newman and Holzman identify epistemology, or knowing, as a paradigm that is in need of replacing. They propose activity (performance) as the non-paradigmatic revolutionary shift that must be taken up (Newman and Holzman 1997, p. 28). Newman and Holzman call for a revolutionary shift away from paradigms to performed activity. Following them in revolutionary, performed activity has reinitiated my own development and has led me to new ways of seeing, being, and creating. Newman and Holzman assert that this unit of study of social therapy is the performance or, in their words, "[t]he unnatural objects suitable for activity-theoretic study are performances" (Newman and Holzman 1997, p. 109).

The "aha" was a sudden recognition that taking a performatory approach to educational transformation is a fundamentally different type of activity from educational reforms, which attempt to solve unresolved problems in the field of education using known methods. The performatory search for method creates developmental STEAM education developmentally. A performatory approach to the STEAM movement creates a new community practice, "a developing development community" (Holzman 1997, p. 65). Holzman's formulation describes a community that is "Vygotskian in its tool-and-result character, for it supports developmental activity and, at the same time, its noninstrumental, nonpragmatic (tool-and-result) is 'merely' the developmental activity it supports (not some other outcome or product)" (Holzman 1997, p.65).

My "aha" was also the realization that my original thought to find a connection between creating developmental environments and STEAM education was in error. My performance of creating this book was developmental. I organized STEAM educators that I knew or met into a community of people who engaged conversations that explored ideas about learning and development in STEAM education. That community grew with each interview and developmentally contributed stories to a book that I hope will support continued development in STEAM learning. The stories I've included about different STEAM learning projects describe how people create development and how they "incorporate the 'other'" (Holzman 1997, p. 73). It turns out that there was no need to find connections; developmental STEAM learning environments will be created by interdisciplinary groups or ensembles.

The interdisciplinary STEAM practitioners and educators I organized and interviewed for this book all indicated in different ways that they look for solutions to problems outside of the disciplinary paradigms they work in. They use approaches that I consider performance-based, although they might not refer to them as such. Newman and Holzman describe what happens in ZPDs as "the social-cultural-historical activity of creating environments for development (zpds)—[is] the everyday, mundane, practical overthrow of existing social relations. People jointly (collectively, socially) transform totalities ('existing social relations')" (p. 109). I believe the key to creating a developmental STEAM learning environment will be in how groups of students and educators transform their relationships to learning, interdisciplinary knowledge, and each other.

The unit of study of my search for method in STEAM education is the ensemble performance, groups of people collaborating creatively. I have provided this brief outline of a few important performatory concepts in the hope of making the ideas throughout the book accessible. The following story is an example of a performatory approach to interdisciplinary STEAM learning that I collaborated on a few years ago.

ZPDs of Math Learning: The Math Video Project

In 2009, I was a middle school technology teacher, and I partnered with a math teacher to create what I called the Math Video Project. The goal was to engage seventh-grade middle school students in math learning in an entirely different way using performance and technology. The math teacher designated Fridays as the project day. The idea was simple: I would take about half of his class out of the classroom every week and work with students for about half an hour. Then, we would switch, and I would get the other half of the class. The project was scheduled to run for six weeks. The math teacher planned to work with small groups and provide students with individualized instruction. I created a program of training for creating the videos that included preparing the students with performance and improvisation exercises. I asked students to create math videos of the math topics that they were learning. I had no concerns about the technology part of the project. I had taught them all. They knew me, and I knew them and what they could do independently. The focus was to work in groups when possible and create videos. The students had a work schedule, technology, the Internet, and each other. They also had my assistance whenever they needed it during and outside of class time. However, I didn't have as much time with the students as I would have liked. The teacher was not as involved with the video project as I'd hoped, and I had no idea what anyone would learn. It was a less than ideal situation.

The seventh-grade Math Video Project resulted in various kinds of videos. Some were Sesame Street™ type creations that were intended to demonstrate basic math concepts. There were a few rap videos and lesson videos that explained how to do particular problems, and there were two videos that we found especially revealing. The first was a music video. A student sang a favorite pop music "breakup" song and set it to lyrics that she and her team wrote. It was a song about how her relationship with a topic in math just was not working. During the project debriefing, the math teacher pointed this video out as being significant to him in that the student's performance changed his ideas about who she was. He had developed the impression that her lack of participation in math class was due to shyness. He admitted that he had misunderstood the student and needed to rethink his approach with her. He went on the describe how, after working on the videos every week, students returned to the classroom excited about the videos they were creating and talked to him about the concepts they were trying to present in the videos. He indicated to me that these conversations revealed misconceptions that the students had about the math topics. He also reported that the assessments he was using were not telling the same story about student misconceptions.

One video that had a lasting impact on me was a solo effort by a student who asked permission to work alone. His video was a digitally animated cartoon teacher explaining the math concept of pi (3.14) to a class of students. The technology he used to produce the video included a laptop with Microsoft Windows ™, the Microsoft Paint ™ drawing tool, and Microsoft Movie Maker ™. He created the script for the voiceover, recorded his voice, and synchronized the recording with the actions of the character in the animation. He probably spent more time creating the animation than most other students devoted to their entire video project.

During the production of the video, he asked for my help with some technical problems he was having. I started asking him questions about pi. He answered with a casualness that indicated to me that he understood the concept. His final project contained a sophisticated combination of technologies, visual art, performance art (he spent many hours perfecting the voice over), and a mathematical concept. I was stunned that he was able to illustrate the depth of his conceptual understanding and apply it in different ways.

Room for Development

Student engagement is what happens when students have the responsibility to make choices and are encouraged to take chances and be creative. I don't try to determine the focus of student projects, but I ask students to pursue interests that are related to the topic at hand. Getting students engaged in an open-ended project with uncertain requirements can produce emotional responses. Emotions range from happiness to feelings of being overwhelmed. I work hard on moving students from being overwhelmed to feeling productive. In the Math Video Project, students made the design choices; the quality of the projects varied, and some design choices revealed more about student content knowledge than others. However, all projects were valued by the math teacher and me. We also publicly shared the student projects with other members of the school.

The math teacher was impressed with the work the students had produced, and he had gained insight about how his students were feeling about math (some videos were celebrations of math) and the ways in which they were struggling with concepts he was trying to teach. He acknowledged that he could not have attempted the project without my initiative. The Math Video Project created many ZPDs that in turn created new performances and development for teachers and students in school.

The following year, the teacher continued Math Video Project without me. I had left the school but remained in touch with him. He had decided that he wanted the students to produce videos on specific math topics. At the time, I found it curious that he wanted to get the students to produce what he assigned. His efforts were instrumental. He wanted the project to produce certain results, although our efforts together had been developmental. Upon reflection, perhaps using an instrumental approach was what he could do, given that I was no longer actively part of his ZPD of creating developmental learning environments. Our collaboration did demonstrate that we could create a performance space in the school day for developmental learning that existed side by side with traditional approaches.

Search for Method in STEAM Learning

I use what I've learned from Newman, Holzman, and others to understand how to teach teachers to use technology and performance in classrooms to create developmental interdisciplinary STEAM learning environments. Given the national push for STEM learning in the K-12 education ladder, I've become involved with people, projects, and schools that are trying to figure out how to integrate STEM and, more recently, STEAM into the school day. I have an undergraduate degree in one of the STEM disciplines (computer science), and I was employed for 16 years as an information technology professional before becoming a teacher. I come to the question of preparing teachers to teach integrated STEM lessons differently than the step-by-step recipe style of many STEM books. I encourage the use of experiential learning strategies, such as project-based learning, hands-on learning, inquiry learning, and service-learning, in teaching practices. Service-learning involves engaging students in topics at school through civic involvement with the community. For example, students might learn about vegetable gardening by volunteering at a community garden.

Through my work with teachers, I have learned how they use scripted STEM curricula and technology in classrooms, particularly K-5 elementary schools. Following scripted lessons is often a mandated strategy for some teachers, but I believe it has some problems. If the ultimate goal of training teachers is to prepare students to "think like scientists or technologists" or "to be creative problem-solvers and innovators" who are "twenty-first-century learners and workers," how will those goals be attained when so many teachers use nineteenth-century approaches to teaching and are not allowed to innovate with the curriculum? Further, how can schools teach students to think like engineers or scientists without having actual scientists or engineers in the room? Finally, is there even such a thing as thinking like an engineer or a scientist? It is easy to imagine that efforts at preparing children to be engineers without engineers in the classroom would be similar to asking a child to learn to speak English as a second language without immersion in interactions with fluent English language speakers.

My interest in finding new developmental interdisciplinary teaching approaches to STEM and STEAM education has led to using service-learning pedagogy to create developmental learning environments or

Vygotskian zones of proximal development in public schools. In the service-learning projects that I have designed, children come into contact with college students and create many ZPDs during project-based learning group activities during the school day.

It is my hope, of course, that this book will help teachers provide STEAM education by creating learning environments that are developmental and interdisciplinary. Chapter 1 provides a context for understanding STEAM education and describes the crisis of the education for workforce development paradigm. Chapter 2 explores the different methodological approaches to STEM/STEAM learning. Part II of the book contains Chaps. 3 through 7, covering the subjects represented by the letters in STEAM (Science, Technology, Engineering, Art, and Math). It introduces the STEAM practitioners and educators interviewed for this book and presents their performances of innovation, creativity, and interdisciplinary practices. Part III outlines service-learning concepts and describes my service-learning research project (Chap. 8). Reflections on creating developmental interdisciplinary learning environments are found in Chap. 9.

References

Holzman, L (2013). Are you a performance activist? Retrieved from http://loisholzman.org/2013/07/are-you-a-performance-activist/.

Holzman, L. (1997). *Schools for growth: Radical alternatives to current educational models*. Mahwah, NJ: Lawrence Erlbaum Associates.

Holzman, L. (2008). Creating stages for development: A learning community with many tasks and no goal. In A. Sumaras, A. Freese, C. Kosnick, & C. Beck (Eds.), *Learning communities in practice*. New York: Springer.

Holzman, L. (2009). *Vygotsky at work and play*. London: Routledge.

Holzman, L., & Mendez, R. (2003). *Psychological investigations: A clinician's guide to social therapy*. New York: Brunner-Routledge.

Martinez, J. E. (2011). *A performatory approach to teaching, learning and technology*. Rotterdam, Netherlands: Sense Publishers.

Newman, F., & Holzman, L. (1997). *The end of knowing: A new developmental way of learning*. New York: Routledge.

Newman, F., & Holzman, L. (2006). *Unscientific psychology: A cultural-performatory approach to understanding human life*. New York: East Side Institute.

Vygotsky, L., Cole, M., John-Steiner, V., Scribner, S., & Soberman, E. (1978). In M. Cole, V. John-Steiner, S. Scribner, & E. Soberman (Eds.), *Mind in society: The development of higher psychological processes*. Cambridge, MA: Harvard University Press.

CHAPTER 1

The Education for Workforce Development Paradigm

Using the idea of a paradigm as a lens for viewing the purpose of education in the USA—a workforce development paradigm—helps make the complex social structure and the limits of policies, practice, and problem domains visible. The way in which the USA approaches STEAM and STEM education is presented here through a review of how leaders and practitioners within the educational community are actively organizing, funding, and training to address challenges of STEAM education knowledge and practices.

The histories of STEAM and STEM are linked, and it makes sense to discuss them together. STEAM is an acronym. It stands for Science, Technology, Engineering, Art, and Math. Some definitions of STEAM indicate that the A stands for art and design. Other definitions suggest that the A stands for architecture. In this book, we will use the definition of STEAM found in a congressional resolution of May 1, 2015, that distinguishes STEM from STEAM: The "innovative practices of art and design play an essential role in improving STEM education and education research," and this is the reason given for adding the A. In addition, "art and design provide real solutions for our everyday lives, distinguish United States products in a global marketplace, and create opportunity for economic growth" (H.R. Res 247 2015, pp. 1–2). Thanks to this resolution, STEAM education is part of the official vocabulary of the US Congress. The language in the congressional record provides educators, and policymakers, with the official rationale for funding education initiatives.

J.E. Martinez, *The Search for Method in STEAM Education*,
Palgrave Studies In Play, Performance, Learning, and Development,
DOI 10.1007/978-3-319-55822-6_1

In August of 2015, the Congressional Committee on Commerce, Science and Transportation submitted a report to accompany the STEM Education Act of 2015 (S. Rept. No. 114–115 2015). The Act added computer science to the definition of STEM, in addition to continuing the support of STEM education programs through the National Science Foundation (NSF). The report states that more support of STEM education is necessary to develop a STEM workforce for manufacturers, high-tech companies, and small businesses across all sectors that struggle to find workers with necessary skills and knowledge to fill in-demand STEM jobs (pp. 1–2). The STEM Education Act became law in October of 2015. Among other things, the Act provided funding for prospective teachers to apply for scholarships, and for the NSF to fund education research in informal learning settings.

Congressional resolutions and various committee reports are how policy advocates, business leaders, and legislators in the USA communicate their views, practices, and understandings of STEAM and STEM education. The previously mentioned documents and many others shape the priorities of government-funded education research along the lines of national priorities that include national security and maintaining global competitiveness in international commerce. Funding education for national priorities is not a new phenomenon or isolated to STEM learning. Early efforts in public education were designed to train children to be "useful citizens" (Rury 2005, p. 3). The primary concern, as stated in the STEM Education Act of 2015 report, is to improve on how the future workforce is prepared to fill "in-demand STEM jobs, including those related to computer science" (p. 2). Proponents of the need for this legislation cite the poor performance of US students on the Program for International Student Assessment (PISA) of 2012. According to the PISA report, American students were ranked as 20th in science and 27th in math among the 34 developed countries that were listed (OECD 2013). These outcomes are considered to be significant problems in the US educational system. The proponents of STEAM and STEM education initiatives link poor performance on international assessments to inadequate preparation for participation in the workforce. This link might not be as strong as proponents of STEAM and STEM learning make it out to be. However, this perception, when viewed through the lens of a paradigm, could be an indicator of an emerging crisis (the inability to solve particular problems in the current paradigm) in the education for workforce development paradigm.

Weak Links

Michael S. Teitelbaum's book, *Falling Behind: Boom, Bust & the Global Race for Scientific Talent* (2014), provides a very helpful overview and analysis of questions related to American competitiveness in the STEM disciplines and workforce demand. In his review of the research literature, Teitelbaum cites numerous government reports and independent research papers. He reveals that there are many stakeholders, such as large corporations and the technology sector, involved in promoting government initiatives in STEM education. Teitelbaum concludes that there is ***no consensus*** among researchers about the preparedness of the US workforce to meet the needs of national interests (Teitelbaum 2014, Chap. 5, Loc 3450 para. 3). His book provides a historical analysis of STEM workforce funding that he describes as "alarm-boom-bust" and reveals the "unstable nature" of government and privately funded initiatives in STEM education and research.

Teitelbaum offers some examples. Both the 1983 report *A Nation at Risk*, published by the US National Commission on Excellence in Education, and the report titled *Rising Above the Gathering Storm*, published by the Academy of Sciences (2007), raised alarms about mediocrity in education and a crisis in global economic competitiveness. The America COMPETES Act of 2007 is an example of "boom" funding, and the US federal government shutdown in 2013 is a case of a "bust" event that unexpectedly constrained discretionary education and research funding at the NSF and the National Institutes of Health (NIH).

In chapter after chapter, Teitelbaum points to the lack of empirical scientific data in government and blue-ribbon committee reports. His analysis and research challenge the certainty of general assertions regarding STEM labor shortages and educational failure. Teitelbaum asserts that despite the limitations of inconsistent federal funding cycles, misalignments in workforce development, and overstatement of workforce needs, the USA is still competitive and produces many students prepared for the STEM workforce. In short, he acknowledges that there are problems in STEM workforce preparation, but "***a real shortage of scientists and engineers is not one of them*** [his emphasis]" (Teitelbaum 2014, Chap. 3, Loc 1879 para. 4).

Teitelbaum points to the work of researchers Lindsey Lowell and Hal Salzman, who analyzed the data that was used to support the alarmist *Rising Above the Gathering Storm*. They released their report in 2007 titled, *Into the Eye of the Storm: Assessing the Evidence on Science and Engineering Education, Quality and Workforce Demand*. In it, they conclude that the reason that the USA lags behind other countries is that the large number of students in the USA impacted by poverty drags down the US ranking in international assessments of science and math performance. According to their report, our best students rank well and with the best students in the world. Lowell and Salzman's report further suggests that we should be concerned about addressing the learning of students performing at the lowest levels if improving the international ranking of students is the primary issue. They point out that there is no evidence that improving student achievement in school will lead to improved national competitiveness.

This selective review reveals that the scientific and education research community, in dialogue with business and government, is responsible for *raising the alarms* and for delivering the *critique of the alarmists.* According to Kuhn, scientists' response to a "crisis" is to identify where the discrepancy is in the field. "The problem is labelled *[sic]* and set aside for a future generation with more developed tools" (Kuhn 2012, p. 84).

Accountability and Achievement

The previous discussion of the education for workforce development paradigm highlighted reports that are used to frame the debate about problems in education. The practices that currently dominate conversations about teaching and learning include measurement of accountability and achievement, standardizing curriculum, and improving the qualifications of teachers.

The No Child Left Behind legislation of 2001 (NCLB) is an education reform that was designed to increase teacher accountability to improve student achievement within the current paradigm. The reports and legislative documents previously cited are responses to the assertions by political leaders, policy analysts, and other experts that education in the USA is in crisis. Common concerns mentioned in the legislation have included the need for more teacher accountability and the need for higher standards (NCLB 2001). Unfortunately, the NCLB reform effort fell short of the stated goals. In 2016, the Obama administration admitted that its revision to NCLB mandates, known as Race to the Top fell short of having the desired impact on reaching underrepresented students in the STEM disciplines (US Government 2014).

Achievement gaps are disappointing to teachers, parents, administrators, and politicians. They are also frustrating, demoralizing, and depressing to students because they are the ones who are coming up short. The problem of student achievement gaps in science and mathematics is another significant concern pursued in the education for workforce development paradigm. The solutions to problems of student achievement have focused on providing more educational funding in the following areas: national curriculum, national curriculum standards, standardized testing, accountability measures, technology in the classroom, increased teacher qualifications, and mandated professional development for teachers. There is education research that confirms that spending more money has helped schools close the achievement gaps between students in poor communities and middle-class students. Baker (2012) provides an example. National educational funding initiatives have supported the participation of more underrepresented students (females and minorities) in the STEM disciplines (US Government 2014). Despite all the money, efforts, and improvements, gaps persist. There is some utility in defining gaps to motivate educational reform. Achievement gaps create a simple way of framing the differences in performance revealed by standardized testing. Policymakers justified distributing local and federal funding to schools with underserved populations or punishing schools that did not make adequate progress by referencing achievement gaps. When NCLB legislation linked the results of standardized testing to criteria for judging the effectiveness of teaching in schools, achievement gaps became a significant concern. Student achievement data was going to be used to determine whether schools were helping students; lack of progress would result in withdrawal of funding and closing or reorganizing failing schools. By "motivating" teachers and administrators with a threat, accountability legislation created the conditions for excessive testing.

Standards

The fourth definition in the Merriam-Webster's online dictionary for the word *standard* reads as follows: *something set up and established by authority as a rule for the measure of quantity, weight, extent, value, or quality*. Standards work very well in manufacturing environments where processes and materials are controllable. The education for workforce development paradigm provides the framework for preparing students to participate in work environments. Measuring student performance is not just desirable but necessary for determining whether or not students are achieving to expectations. It may be helpful to illustrate how standards come into being with an example from mathematics education.

According to its Web site, The National Council of Teachers of Mathematics (NCTM) is a global professional organization of teachers with 60,000 members in the USA and Canada and is the "foremost authority in mathematics education" (Directors 2016). This group is concerned with advocacy, research, professional development, teaching and learning standards, issues of access and equity, and practices. When the reauthorization of laws such as Elementary and Secondary Educational Act (ESEA) is under consideration, the NCTM will produce letters of support for targeted funding for initiatives, such as those related to STEM education, in the reauthorization of the law (Bash 2015). As the experts on mathematics instruction, the NCTM influences national standards in teaching and learning such as the Common Core State Standards for Mathematics (CCSSM). The Common Core State Standards initiative was brought about by the U.S. Department of Education's need to grant waivers in order to continue federal funding to states that were unable to meet NCLB performance standards. One of the priority concerns of the NCTM is bridging the gap between research and practice. The NCTM will organize conferences and appoint committees to develop and publish reports to raise awareness in the mathematics education community. One such report emphasizes the fact that teachers (practitioners) have trouble accessing research and making the generalized findings in the research relevant to their particular circumstances (Arbaugh 2010). In raising awareness about disconnects between new standards, curriculum, and practices that have left teachers and students confused, other stakeholders in the education community can provide their perspectives. For example, teachers' unions and concerned parents pushed back against school districts and state school boards around the country. They claim that the testing is being administered before teachers and students have had an opportunity to adjust to the new curriculum (Weingarten 2013). There is a significant gap between the people who make standards and the people who must meet them. Awareness-raising and collective action are needed to bring practices, standards, curriculum, and theory together.

Don't Reform, Perform!

Many educators will relate to STEAM and STEM education legislation and funding efforts as the latest in a series of workforce competitiveness reforms. They will use the tools they have always used and work on problems in the same ways they always have. We can expect the good and bad results of those efforts to be recognizable as attempts at refining existing ideas about teaching and learning and measuring achievement. I am afraid that the frustration that people experience with education reforms and policy is likely to continue. How could it be otherwise, if the same tools and the same ways of looking at problems continue to be used? A new way of creating change is needed.

Uncritical acceptance of what I have described as the education for workforce development paradigm will make it hard to embrace new ideas and create new practices in STEAM education. The school system works for some students and some teachers, and it does not work for far too many students and teachers. Everyone agrees that more creativity and innovation in schools is desirable; it is in the congressional record. In my experience, thinking of innovation and creativity as something that needs to fit into existing practices is the wrong approach. When innovation and creativity actually happen in an institution or a learning activity, a transformation occurs, everything changes. STEAM educators are calling current teaching practices into question as they create new interdisciplinary practices and ways of being in educational institutions. Their actions, projects, and new relationships are the critiques or the new performances that underscore our need to go beyond reform to achieve/create/realize the transformation of educational institutions that we are all hoping for.

References

Baker, B. D. (2012). *Revisiting the age-old question: Does money matter in education?* Retrieved from http://www.shankerinstitute.org/sites/shanker/files/moneymatters_edition2.pdf.

Bash, E. (2015). National Council of Teachers of Mathematics, letter of support for STEM. Retrieved November 16, 2015, from http://www.nctm.org/uploadedFiles/Research_and_Advocacy/Policies_and_Recommendations/Letter%20-%20Organizational%20Support%20for%20STEM%20Funding%20Provision%20in%20ESEA%2011-16-15.pdf.

Directors, B. of. (2016). NCTM.org. Retrieved from http://nctm.org/about.

Government, U. S. (2014). Progress Report on Coordinating Federal Science, Technology, Engineering, and Mathematics (STEM) Education.
H. R. Con. Res. 247, 114 Cong. (2015).
Kuhn, T. S. (2012). *The structure of scientific revolutions* (50th Anniversary ed.). Chicago: University of Chicago Press.
National Council of Teachers of Mathematics. (2010). Linking research & practice: Executive summary. Reston, VA. *The NCTM research agenda conference report.* Arbaugh, F, B A Herbel-Eisenmann, N Ramirez, E Knuth, H Kranendonk, and J R Quander.
No Child Left Behind (NCLB) Act of 2001, Pub. L. No. 107–110, § 115, Stat. 1425. (2002).
OECD. (2013). Education at a glance 2013: OECD Indicators, OECD Publishing. Retrieved from http://dx.doi.org/10.1787/eag-2013-en ISBN.
Rury, J. L. (2005). *Urban education in the United States, a historical reader.* In J. L. Rury (Ed.). New York: Palgrave Macmillan.
Science, A., & Academy, N. (2007). *Rising above the gathering storm.* doi: 10.17226/11463.
S. Rept. 114–115—STEM EDUCATION ACT OF 2015.
Teitelbaum, M.S. (2014). *Falling behind?: Boom, bust, and the global race for scientific talent* (Kindle version). Retrieved from https://www.amazon.com.
United States. National Commission on Excellence in Education. (1983). *A nation at risk: The imperative for educational reform: A report to the nation and the secretary of education, United States department of education.* Washington, DC: The Commission: [Supt. of Docs., U.S. G.P.O. distributor].
Weingarten, R. (2013). AFT's Weingarten urges moratorium on high stakes linked to common core tests. Retrieved from https://www.washingtonpost.com/news/answer-sheet/wp/2013/04/30/afts-weingarten-urges-moratorium-on-high-stakes-linked-to-new-standardized-tests/?utm_term=.8b1cea4b187e.

CHAPTER 2

Methodological Approaches to STEM/STEAM Learning

What methods and approaches will schools use to train teachers to implement STEM and STEAM learning? The answer unfortunately is: The same methods we have been using for everything else! We can investigate why this is so by looking at how methods and ideas for organizing classroom learning become available to teachers. For the purpose of this discussion, a teaching method is a tool that can be reused to achieve a planned result or outcome. Teacher-preparation programs typically provide new teachers with many opportunities to try out different established teaching methods. The variety of teaching methods available to the profession is beyond the scope of this discussion, but it is safe to say that there is no shortage of access to methods thanks to the Internet. Teachers also benefit from professional development (PD) opportunities provided by schools and school districts. The PD provided by schools figures prominently in how new methods are integrated into teacher practices. School administrators can provide motivation for teachers to take PD classes/seminars/training in the new methods and ideas that a school or school district has decided to budget for. The other way that teachers learn new methods is through additional state-certified professional licensing or through non-degree certificate programs. I can often tell where certain school districts are focusing professional development budgets by the phrases and acronyms teachers use when talking about teaching. One very prominent phrase I've heard over the last 10 years is "student-centeredness."

J.E. Martinez, *The Search for Method in STEAM Education*,
Palgrave Studies In Play, Performance, Learning, and Development,
DOI 10.1007/978-3-319-55822-6_2

Thoughts on Centeredness

A "teacher-centered" methodological approach to creating learning environments features the teacher as the prime motivator of what happens in the classroom. Teacher-centered methods include the lecture, using the blackboard or electronic whiteboard, reading to students, demonstrations, and questioning students. The teacher also decides (as far as the students are concerned) what topics will be learned and how students will learn them. The progressive movement in education and curriculum design has trended away from teacher-centered approaches to student-centered approaches. Student-centered approaches have been proposed as a way to organize teaching in school systems since the late 1890s to address the specific learning needs of students (differentiation) and to respond to low student achievement in schools (Franklin 2005). Most student-centered approaches to teaching allocate the majority of time in a lesson for students to be engaged in cooperative or collaborative activities with peers.

"Centeredness" in learning environments means that there is a focal point around which instruction revolves. In my opinion, talking about whether classroom instruction is teacher-centered or student-centered obscures or oversimplifies the complex cognitive, social, and emotional interactions that teachers and students are having in the classroom. I have heard many educators claim that practice in the classroom is student-centered. However, it is impossible to determine what exactly is going on in the classroom simply because it has been labeled "student-centered."

I see "student-centered" as being a kind of shortcut phrase for describing what happens in the classroom. This shortcut to communicating may be helpful when we do not want to or need to take the time to provide the specifics of student-centered activities. The shortcut does not help when we are trying to find new ways of thinking or innovating in the classroom. What I do think will help is teacher narratives. I've noticed that teachers tend to tell stories about what goes on in classrooms. The stories contain rich descriptions of social interactions in the classroom. Sometimes there are interesting digressions to provide listeners with historical background, and there is often a point being made about teaching in that particular circumstance. These narratives are a genuine and powerful means of engaging adult and youthful learners. To create developmental STEAM learning environments, we are going to have to tell each other stories. In the next section, I will tell an ironic story about learning to use project-based learning (PBL) as a methodology in the classroom. Training in project-based learning has emerged as a popular method for preparing teachers to use student projects as a way to make STEM and STEAM interdisciplinary learning fit into the existing curriculum. PBL training comes with a system of forms and instructions to produce a documented process (unit plans and lesson plans) that will ultimately result in descriptions of student learning outcomes that are tied to explicit learning goals, standards, and products that demonstrate evidence of learning. What follows is an experience in observing and participating in teacher professional development that features project-based learning.

Project-Based Learning Professional Development

During the summer of July 2014, I was invited to attend three all-day professional development sessions with approximately forty teachers in an elementary school. The focus of the professional development was to initiate the creation of PBL unit plans for the upcoming school year. The PD implemented the PBL methodology of the Buck Institute, widely considered the gold standard in PBL training. The trainers were educators who had received Buck Institute training and were very familiar with the schools and school districts the teachers came from. The training was typical of other PBL workshops I have attended. The trainers were knowledgeable and were able to bring computer technology and lesson planning resources to bear that have been shown to be useful in a variety of schools in the district.

Typically, at the beginning of a PD workshop, attendance is taken, teachers drink coffee, eat bagels, and workshop organizers hold off on starting the day until they get close to the expected number of attendees. When that happens, the workshop organizers start making introductions and remind teachers to sign attendance sheets so they can receive what is known as "per-session" training pay. On this occasion, the workshop leaders introduced me as a researcher and a university-based partner. I had an opportunity to introduce myself and speak to some of my priorities, and I took a few minutes to teach and play an improvisation game. The game, "Yes, and" creates a collective story and is designed to help players listen to, accept, and build upon the conversational "offers" that others may contribute in the telling of a collective story. I find that this is a useful game to play when I anticipate being in environments where many people will begin their comments with "No, but" or "Yes, but," which work to negate what has been said and brings conversations to a halt or initiates a dispute. The "Yes, and" collective story is one of my methodological tools for creating developmental learning environments.

The workshop plan was for the participants, all pre-K–5 teachers from three different elementary schools, to work in groups and use instructional technologies, such as laptop computers, the Internet, Google Apps for Education™, to develop STEM-based PBL unit plans. Their PBL plans required identifying a problem and developing a curricular unit that resulted in solutions to the problem. They were required to produce documents using PBL management templates and Web-based resources set up by the school district to provide teachers with easy access. In addition to the materials listed above, teachers also had curriculum maps (a schedule of the content to be taught each month) for the grades they taught and the appropriate Common Core State Standards.

As teachers began to work, I became aware of some resistance to the new ideas and some of the work. Some teachers rejected offers of help. Some teachers seemed to be working on using the PBL framework to retrofit classroom projects. Others appeared to be continuing work started in an earlier workshop. Many of the teachers I worked with had chosen their individual comfort zones as a starting point for a PBL-integrated lesson and were trying to identify a relevant problem to associate with the project unit they were developing. Over the course of the 3 days, even as the teachers became increasingly comfortable with the PBL framework, they struggled to align the standards, curriculum, and ideas. Many teachers experienced frustration at trying to "make it all fit" into their existing understandings of their teaching contexts. I hoped people would remember the "Yes, and" performance when they wanted to say "but," however, many sentences started with the word *but.*

Disequilibrium

According to some of the research literature on teacher professional development, disequilibrium is a necessary component of teacher learning (Opfer and Peder 2011; Wilson and Berne 1999). Existing practices and beliefs need to be challenged for teachers to learn something new. Teachers' responses to the PD were consistent with the research literature. Some teachers demonstrated "resistance" to the experience; I interpreted the failure of participants to make eye contact with the lecturer, their reluctance to ask questions, and their tendency to make statements that began with "but" to be an indicator of this.

Another phenomenon that is identified in teacher professional learning research is that teachers will not adopt new approaches unless they see the benefits regarding improved student achievement (Adey et al. 2004). During the workshop, some of the teachers I interacted with expressed concerns about making PBL structured projects fit within the realities of a school day, meeting the expectations of administrators, aligning projects with standardized testing, and teaching the students. Many teachers who made references to standardized testing said that they could not see how PBL prepared students for the test. Given these conditions, it was reasonable to expect that teachers would continue to resist adoption of new technologies and new methods until they saw the benefits.

Interdisciplinary connections across content areas are part of the natural progression in a PBL unit plan. Teachers with more experience and subject-matter expertise had less difficulty seeing interdisciplinary connections than less experienced teachers. One group of less experienced teachers admitted that they needed to do more research for their interdisciplinary unit on the migrations of native North American peoples. I thought, if the goal of a PBL unit is to generate a process of inquiry, why did teachers feel they had to know the answers in advance? Why could not students and teachers discover things together?

The relevance of instruction to the lives of students is another key feature of PBL instructional units and is one of the objectives of the U.S. Department of Education Magnet Schools grant that funded the teacher professional development at the school I was visiting. In these types of workshops, teachers make decisions about what students will learn based on the curriculum and standards. It was not clear to me how much input students or the community were expected to have in these units. In my interactions with some teachers, it was unclear whether they had an understanding of the socioeconomic realities of the community they worked in or how their social class biases might lead them to take certain things for granted about the lives of their students when making decisions about the relevance of PBL units. For example, one group was planning on having third-grade students create a travel brochure for visiting the Galapagos Islands. I couldn't see how the lesson plan related to the lives of the children in that community, and those connections would still need to be made in the lesson plan, if indeed they could be made.

I observed that experienced teachers seemed to be able to increase pedagogical options in the PBL plans of less experienced teachers, and they seemed willing to share and provide guidance. The beneficial impact of experienced teachers on novice teachers is consistent with some research findings (Adey et al. 2004).

PBL is process oriented, inquiry driven, and presumes an iterative development cycle. The tendency of some traditional approaches to teaching is toward facilitating knowledge acquisition by explaining and motivating students to complete the task. Some units ended with a final assessment of whether or not student-created products met the criteria established by standards. Workshop leaders noted during the workshop that starting the actual hands-on project work at the end of the unit as the assessment instrument was an indicator of teachers' thinking in more traditional terms. The PBL process uses hands-on activities to raise questions throughout inquiry learning units. Based on my observations, it was evident to me that many teachers in the room did experience disequilibrium and were struggling with new ideas. At one point in the workshop, one facilitator did remind teachers of the "Yes, and" story in response to a series of statements where different teachers were saying, "but." It is not the first time I observed someone reaching for an improv method in a moment of frustration.

Dispositions

During the lecture portion of each day, I observed many teachers with "eyes on screens" or who refused to make eye contact with the speaker. That this was frustrating for the trainer was evidenced by the phrase, "You need to pay attention to this." One possible explanation for this behavior is that the teachers were multitasking. I am sure that many workshop participants would claim to have been multitasking. I did see some laptop screens showing e-mails, the PBL forms, and other relevant looking materials. Another explanation, as previously noted, is "resistance," which may be due to indifference, embarrassment at not knowing the material, being unprepared, or being bored. Alternatively, trainers may have mistaken lack of eye contact for lack of teacher understanding. Teacher resistance is a source of frustration in PD environments for trainers and workshop participants alike. The professional development literature helps explain and diagnose teacher resistance, its forms, and possible treatments. But getting to the root causes of the symptoms is not one of the things that can easily be accomplished in a PD workshop.

I engaged in conversations with several teachers and was heartened by their enthusiasm and willingness to plan to take risks with the material. Several of these teachers had already been given formal leadership roles as Magnet school specialists. These were senior teachers who self-selected and interviewed for teaching positions that would be funded through the Magnet Schools grant. These teachers were highly motivated and willing to take on significant challenges, and their performance at the workshop was different from many of the participants. Other teachers were being paid by the hour during the summer to be in the workshop, but their performances told different stories about their reasons for and comfort with being there. I felt that this was a clue to moving beyond describing and diagnosing teacher resistance and toward understanding it.

There are many approaches to providing teachers with support in examining their expectations for students and their beliefs about learning. The best type of support comes from peers and opportunities to reflect openly on teaching practices. In this professional development workshop, there was a plan to provide opportunities for reflection and to use the Critical Friends protocol for feedback. The Critical Friends protocol originated from work at the Annenberg Institute for School Reform at Brown University. It is a type of professional learning community that is designed to structure peer interactions to improve teaching (Moore and Carter-Hicks 2014). The Critical Friends process has a set of protocols, including as a first step the implementation of a "tuning" protocol that provides the group with practice in going through each of the steps in the process together. The outline described by Moore and Carter-Hicks specifies 68 min from introductory activity to closing debriefing (Moore and Carter-Hicks 2014, p. 7). However, circumstances drove workshop facilitators to cut short the feedback and reflection portions (20 min) to cover PBL curriculum development issues. Time for reflection and feedback was traded away for covering the curriculum. I have participated in the Critical Friends protocol and have observed others using it. I view the protocol as a highly scripted ensemble performance. On this occasion, I was an observer, and the interactions seemed a bit rushed. It was hard for me to determine how anyone felt about the process. I do think that the reflection portion is as important or almost as important as the content/curriculum of the workshop.

I think understanding how people felt about the process would (1) help improve the process and (2) probably provide insight into what the take-away for teachers was.

In my opinion, the 3-day PBL workshops proceeded along familiar patterns and would be recognizable as being of high quality despite the varied levels of enthusiasm. The teachers responded along the lines predicted in the literature on teacher professional development. A few days after the workshop, I provided workshop organizers with feedback on the training. The specific feedback is not relevant here; I responded to them with suggestions coming from a best practices perspective. My goal was to continue to build my relationship with these teachers and schools, and that meant I had to work with what they offered, which was an opportunity to provide useful feedback on their terms.

Many teachers feel like they do not have a choice when it comes to professional development, and choices are difficult for PD trainers to create. Empowered teachers, such as those identified leaders (the Magnet school specialists) in a PD workshop environment, will exhibit enthusiasm. The Critical Friends protocols can work when they are routinely part of teacher practices in schools. In my experience, in schools where new ideas take hold, teachers believe there are opportunities for choice making and risk taking. Teachers are also receptive to new ideas if they think that administrators trust them and that they can trust their colleagues. A suggestion I would offer is that schools invest the same effort in creating trusting environments as they do in developing professional knowledge and other professional practices.

Systematic Approaches

Based on my observations of efforts in STEM education, I think that PBL will be the approach that many schools will take toward STEAM education. Collaboration and creativity in classrooms will also be encouraged in STEAM teaching and learning. However, it is still unclear whether creativity and collaboration will be central to STEAM education practices or be viewed as add-ons to what I regard as a systematic approach to learning in schools. Systematic approaches to learning in school sequence and coordinate learning activities. A measurable outcome can be described when the learning process is broken down into distinct steps. For example, "the student will be able to write her name," is a measurable outcome.

When we compare early childhood learning, such as the type that toddlers are engaged in, to formal school-based learning, the differences become apparent. The developmental performatory learning of children outside of school may include, for example, a child's exploration of a living room. The exploration of a room by a child has many possible outcomes, some that are observable and many that are not. The outcomes of an exploration may not be measurable. What a child learns in the exploration of the room may not have direct, causal relationships to what develops and is not predictable.

In a learning activity that is systematic, for instance in a kindergarten classroom, a morning routine might involve children signing into the class by writing their names in crayon on a large sheet of paper. Name writing is re-enforced through the systematic instruction of the alphabet, posting the children's names on personal items, and having them practice writing their names on worksheets and other items. As the school year progresses, teachers will have documented the progress of each child's ability to write her name and form the letters of the alphabet. The expected outcome of instruction and immersion in the production of text is a child who can write her name, recognize letters, and form and space the letters to create words. There is no doubt that a system of learning helps with measuring learning and ensuring that students have opportunities to learn the things that are a priority. However, a systematic approach to learning only recognizes or values the expected outcomes. We cannot discover other important things about children if we only use systematic methods. Fortunately, kindergarten and other elementary school teachers do many things that are, in my view, performatory.

Performing With(in) a System—A Slight Digression

The morning sign-in activity is a non-threatening, formative assessment strategy that is also fun for the students. Elementary school teachers also perform many unsystematic formative assessments of children and their families in daily interactions. For instance, elementary school teachers note how parents and children perform the morning routine. They consciously and unconsciously track changes in the routine, making note of troubling drop-off incidents, children who look sick, or changes in the drop-off caregiver. Any change to the routine may trigger an improvisational response from the teacher. I've known many excellent elementary school teachers who are great improvisers and astute observers of children and families. Those skills and approaches to assessment are performatory and vital to creating welcoming and safe environments for children. In less happy circumstances, where teachers have much less autonomy and do not perform, bureaucratic (systematic) responses prevail, and there is little evidence of development, improvisation, or good conditions for learning. I have worked in hard-to-staff schools, failing schools where the systematic approach to learning dominates, and there are many unpleasant trips to the principal's office. I have had many conversations with teachers about "the system," where they tell me that the system does not allow them to teach much less perform in the ways that I suggest. I encourage them to perform within the system and play with the system. I further remind them that teaching is a political act, and they have a civic responsibility to be advocates for children and families.

Irony and the PBL Workshop

A pedagogical approach like PBL prioritizes what is to be learned and documents it. A PBL may involve many well-defined tasks to produce one or more expected STEAM learning outcomes. However, if PBL outcomes must be predetermined, how will the possibilities associated with unplanned learning be recognized and valued? More important, if PBL and other recently used methods in progressive education are reused for STEAM, would there be justification for expecting different results than those for STEM or other initiatives to improve math and science learning?

I think it is ironic that the PBL method was not the method used to teach teachers in the professional development session described earlier in the chapter. Professional development workshops are product oriented. Teachers must produce unit plans for teaching, and the workshop is a process for production, not a process that prioritizes inquiry or facilitates the involvement of stakeholders (members of the community, students, etc.) in the development of the unit plans. For teachers, learning the PBL method can get disconnected from practicing the method. To be sure, many teachers do produce PBL units that are engaging and efficient in this manner. However, I question the sustainability of this approach. The PBL system generates a significant amount of documentation that details what students need to do, how activities will meet standards, and how student performance will be assessed. Unit plans also include listings of required materials, interdisciplinary connections, differentiated strategies, and expected outcomes. Teachers will tend to reuse and perhaps revise units, but what will occur when there is a change in the curriculum or the standards? What will happen when a second-grade teacher is reassigned to teach the fourth grade and her PBL units are no longer relevant? Will she be offered someone else's fourth-grade PBL units? Will she find them appropriate for how she envisions teaching the fourth grade? What will happen when funding for teacher PD and new curriculum development efforts ends? A challenge of having any system is that it needs to be maintained and moreover that it can break when conditions or assumptions change. Another challenge of systems is that they encourage more systems, which can lead to fewer opportunities for creativity and autonomy.

Despite my questions about the PBL approach, I believe it is possible to use systematic approaches in creative ways. We can play and perform with the system and within the system if we need to. The value of project-based learning is that it does provide students with hands-on learning experiences. When a PBL unit is ambitious and well-designed, there are opportunities for collaborative learning experiences inside and outside of the classroom with peers and adults.

Experiential Approaches

Project-based learning provides a type of experiential learning. Experiential learning can include but is not limited to field trips, collaborative research projects, internships, service-learning, and study abroad experiences. Descriptions of experiential learning do not usually include imaginative play, rule-based play, team sports, improvisational performance, theatrical performance, and organizing public exhibitions. I believe the play and performance activities that I've added to the list are all forms of experiential learning that should be part of any approach to developmental STEAM education. Experiential approaches to learning provide students with opportunities to reflect on what they are doing and learning. The reflective process of experiential learning can be about more than generating a piece of writing that will be submitted at the end of a lesson. Reflection can be a part of an ongoing process that informs creative development. What I find most powerful about experiential approaches to learning is that they often take place in a "real world" context. When the outcomes are not overly predictable or predetermined, students must bring the entirety of their being to bear on figuring out what they need to do, not just report on some knowledge they acquired. However, even experiential approaches to learning can be made to be as systematic as any other kind of approach. What makes one approach to learning systematic and another unsystematic or performatory?

The Math Video Project discussed earlier was designed as a developmental, performatory approach to learning. I could not predict the outcomes, and I did not predetermine what learning standards would be met. Furthermore, I couldn't claim that I "knew what I was doing" because I had never done it before. I was confident, however, that something positive would come out if it because students were being supported to collaborate, they were using new tools, and they had complex challenges that were relevant to their lives.

If I were to make the Math Video Project systematic, I would determine specific content knowledge to be covered by all videos. For example, using seventh-grade math content, the theme of the videos might be to understand the concept of pi. Each video would have to meet criteria that aligned with learning standards in mathematics and presentation skills. Each team member would be assigned specific roles in the project and would be responsible for specific tasks. There would be a test at the end of the production of videos to confirm that everyone learned something about pi. I would still expect to get a variety of videos, but they would all be about pi. The students would still have opportunities for choices, and they might still have fun and be engaged because they are using technology.

I do know that the overall experience would be different because I have done projects with students using performatory developmental approaches and systematic approaches. Students and teachers can become very comfortable with systematic approaches to learning because they know what to expect and what is required. Knowledge is acquired incrementally, and as long as a student does not fall behind, progress is predictable and measurable.

When I have used performatory approaches with middle school students, I upset the order of things. Students will ask questions about the requirements when they do not see many. They will express uncertainty about whether they are doing their projects correctly. Students will often discover that certain approaches to a project can lead to dead ends. Students tap into their personal areas of strength, and some discover things about themselves that they would like to improve. Many students are often more self-critical about their performances than I would ever be of them. A performatory approach to teaching is more fun and interesting, and it creates opportunities for different kinds of wonderful conversations with students. The conversations that I have with students contain feedback that they can use to continue to develop their performances. I also build better relationships with students when I use performatory approaches. Experiential learning, especially when there are opportunities for "real world" interactions, creates development in many of the same ways that a performatory approach would. Experiential approaches to learning help create stages for performatory approaches to learning and development.

References

Adey, P., Hewitt, G., Hewitt, J., & Landau, N. (2004). *The professional development of teachers: Practice and theory*. Dordrecht: Kluwer Academic.

Franklin, B. M. (2005). Progressivism and curriculum differentiation: Special classes in the Atlanta public schools 1898–1923. In J. L. Rury (Ed.), *Urban education in the United States, a historical reader* (pp. 119–135). New York: Palgrave Macmillan.

Moore, J. A., & Carter-Hicks, J. (2014). Let's talk! Facilitating a faculty learning community using a critical friends group approach. *International Journal for the Scholarship of Teaching & Learning, 8*(2), 1–17.

Opfer, V. D., & Pedder, D. (2011). Conceptualizing teacher professional learning. *Review of Educational Research, 81*(3), 376–407. Retrieved from http://doi.org/10.3102/0034654311413609.

Wilson, S. M., & Berne, J. (1999). Teacher learning and the acquisition of professional knowledge: An examination of research on contemporary professional development consulting editors: Deborah Ball and Pamela L. Grossman. *Review of Research in Education, 24*(1), 173–209.

PART II

Dialogues

Interdisciplinary Development

I expect that interdisciplinary project-based learning (PBL) will be a part of STEAM learning in schools. As a result of doing the research for this book, especially in the interviews with interdisciplinary educators, it became apparent that the separate chapters representing the letters in STEAM did not completely reflect how practitioners were doing science, math, and the other STEAM subjects in interdisciplinary learning and teaching activities. For example, in Chap. 3, scientists who presented at the Cultivating Ensembles in STEM Education Research Conference spent more time talking about using the performing arts in teaching than in talking about traditional ways of teaching science. I think interdisciplinary practitioners will follow the problems that are not solved within disciplinary silos into the next discipline and the next new practice. The STEAM movement contributes to discovering different approaches to learning, teaching, or content-specific practice. What I learned about interdisciplinary practitioners was that they were creating interdisciplinary zones of proximal development (ZPDs). What follows is a deeper consideration of performatory and developmental interdisciplinary STEAM settings.

Learning and Development

In the introduction in Part I, Vygotsky and his discovery—the zone of proximal development—were introduced. I use the words *learning* and *development* frequently, and when used in the Vygotskian sense, they are not synonyms. To Vygotsky, "good learning" happens in "advance of development" (Vygotsky 1978, p. 89). A brief illustration of everyday learning and development may help to bring "learning" and "development" into focus:

Matthew is 14-months old. He is developing at the astonishing rate that is typical of babies. His babbling sounds like the words that adults use in their interactions with him. Based on his alertness, willingness to interact, and the constant smiles, he can be considered a friendly and happy baby. He recently achieved a developmental milestone. He has transformed from a crawling baby into a toddler who has learned to walk. Running, jumping, climbing, and reaching are all new developmental possibilities for Matthew, much to his mother's concern. In his wake, Matthew leaves a trail of toys littered about rooms and brings the possibility of destruction to glassware, floor lamps, electronics, books, and magazines wherever he goes. Of course, Matthew is not intentionally destroying anything. New mobility works to develop his shifting interests in the things that he can see, touch, smell, taste, and hear. Matthew has learned to move, and now objects of interest are within reach. He is a little storm of learning and possibility. For Matthew, crawling is a development of the past, something he will do less and less as he continues to grow. These new developments and the learning that leads it are the foundations for future learning and development. Matthew has learned to walk, and that has changed everything about Matthew, his relationships to his environment, and the people in it.

Developmental learning is the type of transformation described above. Just as Matthew has learned to walk, in part, by imitating others who walk, similar changes are simultaneously occurring in Matthew's language development, which, among other things, his new mobility has created. There is more urgency in communications with parents and other caregivers. Matthew's new range of movement has created more physical interactions. Matthew's caregivers call his name with much more frequency to draw his attention, and they use language to get him to stop or pause before the next mishap. It is in these and many other social interactions that Matthew and his caregivers will create meanings of words. By using language, his caregivers relate to him as being "a head taller" than his current level of development (Vygotsky 1978, p. 102). He hears the sounds that others make when speaking words. He will recreate the sounds he hears with his efforts to use his lips, tongue, and vocal chords. He can respond to words, but he does not understand the meanings in the same ways that experienced language users do. Not understanding precise word meanings does not prevent Matthew from "imitating" experienced speakers by performing speaking and interacting (Vygotsky 1978, p. 87). Whether or not Matthew understands what the adults are saying does not seem to matter to the adults either. They are performing speaking with Matthew. They are not using "developmentally appropriate" vocabulary that will be in the school curriculum in a few years when he enters school. Typically, school-based approaches to learning do not relate to children at the "head taller" level that is necessary for developmental "good learning" that Vygotsky specified. In schools, learning is separated from development or, worse, mistaken for development (Holzman 1997).

According to Newman and Holzman, Vygotsky observed that children do not imitate everything, only what is in their ZPD (Newman and Holzman 2014, p. 70). They view Vygotsky's claim as significant to what it says about language acquisition as an activity.

> Imitation in the ZPD is the activity of meaning-making, where the predetermining tools of the adult language and the resulting predetermined tools of mind are used by the child—the toolmaker—to create something that is not determined by them (Newman and Holzman 2014, p. 70).

They go on to elaborate on meaning-making by making the following distinctions.

> The child's imitations, in contrast, are not determined by the predetermining tools; they are the use of such tools for results to create tools-and-results, to create meaning and, thereby, to reorganize thinking/speaking. While we do not know what the child means when she/he imitates what is proximal to her/his development, we do know that the child almost certainly cannot mean what the adult means (Newman and Holzman 2014, p. 70).

They conclude the elaboration of meaning-making by highlighting the unity of activity and using predetermined tools like language.

> It follows, then, that what we know—and this is most important—is that the child means, because for the child meaning is not yet separated from the total activity of meaning-making, as it becomes for the more fully alienated (societally adapted) adult (Newman and Holzman 2014, p. 70).

The introduction of this book links together development, performance, and play. Play and performance are developmental, and when a child plays, she is "a head taller" than her current level of development. Imitation and meaning-making are developmental activities that happen in play. These are not the activities that are encouraged in school-based approaches to learning and instruction that is aimed at a child's current level of development. To create developmental STEAM learning environments, there must be opportunities for imitation, meaning-making, and performing "a head taller."

Meaning-making and STEAM

The learning done in school is for the most part non-developmental; it does not produce transformations of the type we see in the home in early childhood language development. Schools support well-defined, curriculum-specific opportunities for learning that are designed to result in the incremental acquisition of knowledge and skills. In learning without development, a child passively waits for someone else to present the next thing to be learned. Anyone who has spent time around toddlers recognizes that they are not waiting around for instruction. By doing, moving, and performing, they are learning and developing.

Lois Holzman frames her Vygotskian description of learning and development as performance. According to Holzman, human beings are in a dialectical relational process of actively and simultaneously performing "who we are" and "who we are not" (Holzman 1999, p. 52). Holzman and Vygotsky are both describing Matthew's development in a new way: a baby performing as a speaker before he can speak. In a ZPD, what is called the process of becoming is a process of transformation in relationship to others and the environment. Holzman and Vygotsky view ZPDs as encouraging performances and shared activities that are in advance of what members of the ZPD already know how to do.

In the final paragraphs of *Mind in Society,* Vygotsky shares some conclusions he has drawn from his research and the implications for formal schooling. Among them "'writing must be relevant to life'—in the same way that we require a 'relevant' arithmetic" (Vygotsky 1978, p. 118). Vygotsky also thought that writing should be meaningful and that it should be "cultivated" rather than imposed. He criticized the mechanical way that children were taught to write and learn other skills in the schools. Vygotsky explored the value of instruction in leading development across various topics including reading and writing, grammar, arithmetic, natural science, and social science (Vygotsky 1987, p.179). Vygotsky's research reveals that writing is difficult for children because it requires levels of abstractions that disengage the child from language as a sensory and relational experience.

Newman and Holzman build on Vygotsky's ideas on children's writing: "...drawing and play should be preparatory stages in the development of children's written language" (Vygotsky 1978, p. 118). They write, "[i]t is children's play with written language that makes it possible for them to learn, eventually, the 'workings' of written language" (Newman and Holzman 2014, p. 88). Vygotsky and Newman and Holzman are referring to the early stages of development in writing when we praise children for the pictures and marks on a page that are their writings. We praise a child for the stick figure that stands larger than a house on a page. It is not unusual to see a larger stick figure holding stick figure hands with a smaller stick figure. The labels of the stick figures might say "daddy" with the *d*'s written backward, and the child's name might have other creatively oriented letters. The praise we offer is how we are relating to the child as a writer. As far as her parents are concerned, the child has authored an autobiographical account of her life and relationships. In school, the child might be related to someone who is learning to write. As such, the child will get feedback designed to correct the writing. School-based feedback that corrects writing changes the writing activity. Accurate reproduction of symbols and organizing text will take precedence, and the meaning-making characteristic of creative and playful writing activity recedes. The teachers in school relate to the child's writing as if she were not meaning-making with written language yet. Recall that Matthew's early language development in the home does not require him to use language correctly or intelligibly for him to be related to as a speaker by the adults in his life.

If students and teachers are going to discover how to integrate STEAM education into school in a developmentally meaningful way, then we are going to have to move from abstractions that disconnect learning from meaning-making and start playing and performing STEAM learning. Children are not the only ones who learn in play. Adults can also develop through play and performance activities. I've learned to be creative by trying to create learning environments that are creative and developmental. I've learned performance and improvisation games in workshops with adults that have been very useful in my efforts to reinitiate my own development. If educators are going to use new tools such as performance, then they must reinitiate their capacities for developmental learning and performance. It should be noted that there is nothing trivial about asking adults who have substantial investments in approaches to learning that they have had success with to change their approaches to teaching and learning. However, I do believe that changing our personal approaches to learning will be necessary if we want to create developmental STEAM learning environments.

A casual search of the Internet using "STEAM education" as a keyword phrase will result in thousands of Web sites with STEAM-related lesson plans with projects and products to encourage STEAM learning. These Internet-based learning materials will be helpful, but they will not create developmental learning environments. Scripted learning materials do not contain the developmental meaning-making and ways of relating to children as "a head taller" that are necessary for development. The interdisciplinary character of STEAM education means that new concepts and new approaches to learning content will be part of a developmental STEAM learning environment.

Play and performance are powerful ways to learn new things developmentally. As a person who has been learning to use new technologies for that last 30 years, one of the first things I do with new technology is play with it. Playing with technology as an adult means that I use the technology for non-serious reasons before I decide to invest time and effort in it. I discover what it is supposed to do and how I am expected to use it. I give myself time to get a feel for the user interface. I try it out with other technologies that I use, and if I am lucky, I will be able to include others in playing around with the technology. If I still like the technology at the end of playtime, I will probably start using it for work purposes. If I don't continue to use the technology, then I've learned, in play, something about my preferences, and I'll probably remember the experience the next time I consider a similar piece of technology. I expect that educators who are teaching in technology-rich interdisciplinary environments will have to spend a lot of time playing with technology.

I anticipate that interdisciplinary zones of proximal development will have different types of technologies that students and educators will be learning to use to create developmental STEAM learning environments. These new technologies will transform what we currently think of as classrooms. I also imagine that anyone might be a contributor or a builder of an interdisciplinary ZPD, that anything might be learned in a ZPD, and that the student and teacher roles will become blurred. I think that communication in the interdisciplinary ZPD will be interesting and challenging because in creating the ZPD we will be taking on new meaning-making activities. Each discipline in an interdisciplinary ZPD may contain a specific approach to language and concepts that may be new to each member of the ZPD. In an interdisciplinary ZPD, there will not be time for everyone to be trained in every disciplinary domain that is present. Each member of the group or team will have to learn to relate to others as "a head taller" as they perform "who they are not" with a new language, tools, and concepts. Relating to learners as being in a process of becoming, as opposed to being deficient in knowledge, will be something that is very new and different in academic settings.

References

Holzman, L. (1997). *Schools for growth: Radical alternatives to current educational models.* Mahwah, NJ: Lawrence Erlbaum Associates.

Holzman, L. (1999). Life as performance. In L. Holzman (Ed.), *Performing psychology: A postmodern culture of the mind.* New York: Routledge.

Newman, F., & Holzman, L. (2014). *Lev Vygotsky: Revolutionary scientist: Classic edition.* Abingdon: Psychology Press.

Vygotsky, L., Cole, M., John-Steiner, V., Scribner, S., & Soberman, E. (1978). *Mind in society: The development of higher psychological processes.* In M. Cole, V. John-Steiner, S. Scribner, & E. Soberman (Eds.), Cambridge, MA: Harvard University Press.

Vygotsky, L. (1987). In A. Kozulin (Ed.), *Thought and language* (2nd ed.). Cambridge: Massachusetts Institute of Technology.

CHAPTER 3

Science

Labels simplify things, and the label "scientist" simplifies our concept of the lives and the work of people who work in a place called a laboratory. Whatever images a scientist in a laboratory evokes, it is certainly not the entire story of the person who is a scientist and what his or her work means. Over the past few years, I have had the opportunity to work with some scientists who are interested in performance, first as an attendee at a conference and then as a member of the conference organizing committee at a subsequent meeting. What I've learned is that scientists do not just work in laboratories and they have many ways of thinking about science and the world. In this chapter, I describe the conference where I met scientists who are doing interdisciplinary work using the arts in teaching science. A conversation with an architect who plays with the building blocks of materials and biology will help with understanding different aspects of interdisciplinary work in teaching the STEAM disciplines.

Interdisciplinary Problems

In the twenty-first century, the idea of a field in science is not as simple as biologists working on biology. For example, what does a biologist who specializes in computational biology do, or what does a biophysicist do? There is a general understanding that science is becoming interdisciplinary. This idea is supported by evidence derived from analysis of scientific publications. Porter and Rafols found an increase in the number

J.E. Martinez, *The Search for Method in STEAM Education*,
Palgrave Studies In Play, Performance, Learning, and Development,
DOI 10.1007/978-3-319-55822-6_3

of scientific journal articles that were interdisciplinary (Porter and Rafols 2009). Interdisciplinary science is also a topic of interest in academic institutions currently redesigning educational programs to meet future workforce needs. In 2015, the weekly science journal *Nature* published a special issue on interdisciplinarity. The issue covered topics ranging from the need for interdisciplinary teams to take on the world's biggest challenges (Ledford 2015) to the challenges of funding interdisciplinary research (Rylance 2015). Proponents of interdisciplinary approaches to scientific research argue that the complex problems of the world require teams working together collaboratively. According to Rylance, the critics argue that interdisciplinary approaches are still dependent on distinct disciplines and that interdisciplinary research is a distraction that drains resources from high-quality research activities. Assuming that both positions have merit, how will educators prepare students for work in the sciences? There are no simple answers to questions about what a scientist does.

A different way to think about how students come to participate in the sciences is by taking a brief look at routes that people have taken to a science-related profession.

> My friend Dr. R. specializes in cosmetic dentistry. His motivation for becoming a dentist came from a field trip that he took while in high school. He had the opportunity to watch a dentist transform an old woman's face by giving her dentures. He thought, "I can do that!" Dr. R. was also good at painting, and in high school, the advice he received from his guidance counselor was to take up a vocation like house painting. Obviously, he did not heed that advice. My friend is of Jamaican descent, the first generation born in the United States. He cleaned toilets to pay his way through college. Today he is the owner of a successful dentistry practice, a generous contributor to his community, a teacher at a school of dentistry, a husband and a father putting children through college.
>
> Dr. G never intended to become a dentist. He grew up in California in the 1960s. His interest in college was in anthropology. Then he met a girl who was studying dentistry and his relationship with her exposed him to a new area of interest. He changed his focus and began his training in dentistry. His relationship with the girl didn't last, but he has had a long career in dentistry, much of it spent working in non-profit community settings providing dental care for the Native American community in San Francisco. He has won lifetime achievement awards and has authored books and scientific research articles on TMJ (temporomandibular joint) disorders and

> the use of acupuncture for managing pain. He even authored a book on salmon fishing. Dr. G is retired now, but he often consults with the government on cases related to forensic dentistry. He continues to travel to scientific conferences around the world.

The science and the practice of dentistry do not capture the fullness of the lives of the men presented in the anecdotes above. While they are both dentists, they come to dentistry with different histories and approach the practice in different ways. Their trajectories were not predictable, and their work in dentistry is not separate from the lives and interests they have outside of the dental office. What is worth noting here is the interdisciplinary nature of their interests: Dr. R. is interested in aesthetics and visual art, and Dr. G.'s interests range from anthropology to salmon fishing. I've discovered that personal interests tend to be inseparable from interdisciplinary professional practice. If educators are going to succeed at encouraging and inspiring students to consider STEAM careers, starting from student interests and their own might be a promising approach.

Cultivating Ensembles in STEM Education and Research

The first Cultivating Ensembles in STEM Education Research (CESTEMER, pronounced Keh-STEM-er) conference in 2011 was a meeting on performance, science, and science education, hosted by the University of Connecticut through an NSF (National Science Foundation) research grant. The grant funded a research project titled, *Improvisational Theater for Computing Scientists.* Abstracts from the conference are available at http://improvscience.org/cestemer. Dr. Raquell Holmes, who is a computational cell biologist, and a pioneer in the use of improvisation and performance in developing science research communities, organized and led the conference. Dr. Holmes is also the founder of improvscience, a professional development organization that uses improvisation and performance to help scientists collaborate and build research communities.

Conference presentations focused on how performatory modalities, such as theater arts (improvisation, dance), craft arts (crocheting, hair braiding), visual, and media arts, could function as methodological approaches to learning science and math in higher education and K-12 settings. Some conference presenters were early career scientists using performance in their teaching, and others were well-established practitioners at well-known institutions who were looking for ways to be more inclusive of diverse communities of students and colleagues in interdisciplinary STEM education and research.

The CESTEMER conference experience is different from the large academic conferences that I usually attend with thousands of attendees. The number of attendees at the first two conferences has been under 100, but we did grow from that first conference in 2011 to the 2015 conference in San Francisco. We expect further growth for the 2017 meeting in Chicago. Another difference between CESTEMER and other conferences is in the activities of the conference presenters and their interactions with attendees. Accepted conference proposals feature hands-on activities, and many include learning theater performance games, movement, arts and craft activities, and making games out of science content. Presenters will often include video recordings of activities with students in science classrooms and invite the conference attendees to try the activities themselves. During the breaks in the schedule, large ensemble performances and collaborative activities are organized.

The small size of the conference provides the organizing committee members with opportunities to foster personal connections with conference attendees. Attendees can connect with each other during meal breaks and after the conference when the day's scheduled events are over. I felt so welcomed and included at the first conference that I did not hesitate to join the organizing committee for the second conference and the third. The playful and performance-oriented vision of the meetings created opportunities to experience science learning from new perspectives. Attendees felt that ideas and experiences at the conference could be implemented back at their institutions. For many people at the conferences, the theater games were unfamiliar. However, the briefly shared anxiety about being invited to perform dissipated after the first couple of improv games during the opening plenary, and most attendees began to relax and enjoy a new way of being at a conference.

Student engagement in the sciences is a major focal point at the CESTEMER conference, and many of the presenters and attendees are science educators. They recognize that getting underrepresented students involved in science requires finding different ways for students to relate the content and practices to their lives. Another area of CESTEMER focus is the development of the so-called "soft skills" through performance. Increasingly, research scientists find themselves working on interdisciplinary projects that bring them into contact with colleagues in other scientific disciplines, the social sciences, and the liberal arts. CESTEMER attendees and conference organizers recognize that working across disciplines and creating shared understandings is very challenging. Improvisation, theater games, and the visual arts help with developing listening skills and creating new ways of meaning-making across disciplines.

In a video-recorded interview at the 2011 conference, Holmes explained her vision of performance and science (http://improvscience.org/cestemer). She wants people to understand themselves as performers and believes this understanding will give them the opportunity to create conditions to change their work environments and their lives. Her hope is that people at the conference will engage in a new dialogue of performance, where scientists and science educators can develop relational skills and create "the stage" of science where they are growing as performers and educators.

An Interdisciplinary Conversation

My university, like many others, is undergoing much-needed changes in response to the needs of students. One happy consequence has been an influx of new colleagues. One, in particular, Christian Pongratz joined our faculty from Texas Tech University. At Texas Tech, he was the Director and Founder of the Digital Design and Fabrication Program, which is part of the Master of Science at the College of Architecture. Among his notable accomplishments, Pongratz is the co-author of a book *Digital Media for Design* (Perbellini and Pongratz 2015) and also teaches at several institutions worldwide in the continuing education program of the American Institute of Architects. Before establishing his practice, he worked in New York for Peter Eisenman and John Reimnitzand, was involved with international invited design competitions, and with the design of prestigious commissioned buildings. Pongratz is an educator with a global perspective and has received awards for his work in interdisciplinary scholarship and teaching.

Pongratz presented his ideas about teaching in interdisciplinary areas and architecture at a faculty luncheon. Many of the ideas were new to me, and some sounded like science fiction. His talk was about the intersection of architecture, fabrication technologies, information technology, biology, and physics in the architecture design laboratory environment.

According to Pongratz, students in architecture use digital information to drive the latest fabrication equipment. Laser cutters, 3D printers, and CNC (computer numerical controlled) routers are tools that take a digital diagram as input for cutting materials, as in the case of laser cutters and CNC routers, or extruding materials, as in the case of 3D printers. All of these tools are capable of rendering two-dimensional or three-dimensional objects. Laser cutters, 3D printers, and CNC routers are the latest "must have tools" for design studios and fabrication laboratories.

Pongratz believes individuals will be able to create solutions to problems on a small scale using digital fabrication tools. Cities will become places where high-tech manufacturing is carried out, and cities will import digital products from the digital information stream that will be used to drive the fabrication tools that will render products in final material form. From his perspective, small-scale fabrication tools have implications for the design of cities. If, for example, manufacturing no longer happened on the outskirts of cities, commuter travel between home and work would change.

With digital tools, it will be possible for biology to become part of the design and fabrication process; biological cells would be just another kind of building block. Pongratz describes nanoscience as enabling the design of products that are at the scale of a strand of DNA (deoxyribonucleic acid). According to Pongratz, it is just a matter of time before design happens at the level of atoms. Pongratz is excited about what he sees as the potential of the small-scale fabrication Maker movement, which enables anybody to get involved in making and designing things. At his talk, I learned that architecture was about more than designing buildings and cities. Pongratz sees it as a practice in which the disciplinary boundaries have become blurred. A day after his talk, I e-mailed Pongratz and invited him to be interviewed for the book I was writing. He agreed, and what follows are some edited excerpts from our conversation.

Jim: How did you become an architect?

Christian: I was interested in biology and chemistry when I was in high school. Another direction was my Art class. I was very attracted to both of those areas. When I reached the university level, I started with computer science. I didn't like it, and I moved to architecture.

Jim: I thought your ideas about architecture and interdisciplinary learning fit surprisingly well with my ideas about human development and learning.

Christian: The profession of architecture, by default, is dependent on many others. Speaking for myself, you really can't do anything alone. You need input all the way through, as early as possible so you can understand potential implications down the road. It's not like you do something, you hand it off to the next person, they do something with it, and it goes through a chain. It's all a creative interplay between what everybody thinks about the process. If we go back to the building blocks that you mentioned from my presentation, I believe that this is the whole story. In fact, for me, the building block of an atom and the building block of a neuron or a molecule were always the same thing.

Jim: How are they the same?

Christian: There is not a difference, once you zoom down small enough to that level. Once one understands the organization of those substances, and that there is an underlying system somewhere, one can start to play with that. The beautiful thing is that it's scale-less. I try to communicate that there is a large scale that is traditionally understood as being architecture, and there are many other levels that were always architecture, and they are now becoming parts of the architecture.

Jim What are your views on the Maker movement?

Christian: I'm glad I had the chance to mention that in my talk because I am terribly fascinated by it. When we look from an industrial Engineering standpoint, factories are designed with very precise organizations and systems. This Maker culture is a sub-culture, an anti-movement from people against something that is prescribed. This movement doesn't follow any pre-set organization, as science would teach you. Makers learn by playing with something and asking somebody for assistance. But, the fabrication machines are not necessarily used by Makers in the way they were designed to be used. What is amazing is that we are tapping, if you think on a global scale, into a reservoir of creativity of people who were never asked to apply their creativity to those types of objects, be it a soldering station or a 3D printer. In product development companies, the way they go about developing products is through a pre-set process.

Jim: What is that process like?

Christian: There's a designer, there's some marketing guy, there's some producer, and they follow pre-set systems, and they are probably successful. They bring in creative people to discuss the implications of "what if?" scenarios. But, they don't have anyone involved in the process that is not from the field. What is new in the Maker movement is that anybody can make things that are traditionally produced in factories with a pre-set process.

Jim: How are Makers different from other product creators?

Christian: The Makers may have no idea what route, what process they might want to use. They go a non-traditional route. They will tap into areas that will reveal some new knowledge. On the one hand, this has a disruptive effect, and we don't know what it will do or whether it will work. On the other hand, the process of creating a new product that could fail may even be more valuable than what you could extract from a regular pre-set process. It could lead to something better. There's a guy, Richard Florida, who was at Carnegie Mellon. He talks about the creativity of everyone. We just haven't had a system to benefit from that. In my mind, that is the system that is emerging right now. The Maker movement is what everyone can tap into, and then we release that creativity that everyone possesses.

On the one hand, Pongratz describes a scientific manufacturing process based on a manufacturing domain of knowledge. Manufacturers and product companies invest in the processes and knowledge of that domain. They operate within the manufacturing paradigm and use the paradigm's tools. On the other hand, Pongratz is saying that people or Makers who do not have domain knowledge are using similar tools and unproven processes to create products—and methods—that may be outside of what is generally acceptable in the manufacturing paradigm. In doing so, Makers disrupt everything we thought we knew, and while they may fail, we may learn new things from those failures that would not have come into being using the methods of the established paradigm. The words that Pongratz is using, like the word *system*, is scientific, implying domain knowledge and practices. I would argue, however, that the disruptive creative process that Pongratz is describing is unsystematic and performatory. He is saying that knowledge of systems is not a prerequisite to using tools or inventing new methods. From my perspective, this is similar to when children learn to use language, don't know language is a system. The Makers are performing as manufacturers in a new and interdisciplinary way; and they, too, are learning, developing, and creating. The Maker performance is not just a manufacturing performance, but it is a creative activity that defies fixed labels. Everyone performs as a Maker, whether it is making a meal or making some artwork, the Maker movement gives people an alternative to being consumers of products and/or knowledge.

Jim: The opportunity that I recognized in your talk is that all of the knowledge domains are up for grabs, the systems and the disciplinary silos are disrupted.

Christian: Exactly. And we don't know yet which disruptions will be created. With the Internet, we learned that during the mid-90s to the early 2000s it was a free environment, but you didn't have everyone participating yet. I could not yet send e-mails to my friends in Germany because they didn't have Internet access. The big corporations at the time didn't know if they should have a Web page or not. Barnes and Noble still sold the books in the regular shop; they didn't go online until late in the game. Now, in small-scale manufacturing, we are exactly at the point where there is no big corporation in this new movement. Of course, if the small-scale manufacturing movement steps on the toes of too many corporations, it's clear it will all change. We will not produce everything like that, but individual elements and products we can produce on a small scale.

Jim: Who will be the early adopters of these new methods of production?

Christian: The younger generation, who are very into gadgets and everyday playful objects, skateboards and stuff like that. They will do this. Students might come up with an idea to make something better than what they find on the market. The products they come up with will at least be customized. A product will have the student's name on it, a drawing, something that is coming from the student's personalized input. That's a big driver. We can certainly make things on our own, and people can produce in areas where making solves needs.

Jim: What's your global perspective?

Christian: In Africa and Eastern Europe, people can benefit from a creative group that is somewhere else on the planet. The group becomes aware of where a need is, and they can think of a larger picture, and find a solution. The idea is transmitted on a digital stream, then it's locally produced. It solves the local need. You have a global knowledge transfer paradigm and an extremely localized need for production. That's an amazing thing.

Jim: Where do you think the community engagement piece of the Maker movement goes?

Christian: The fab lab, a small-scale fabrication laboratory. You can place it in whatever community or whatever place in the world. You don't care who is living there, what the culture is, or what race they are. The technology might be more advanced than whatever household machines people have there. I think they will sit down, and they will make something.

A similar idea, dropping advanced technology into a rural setting and watching what happened, is what Sugata Mitra's work in education in remote communities in rural India explored (Mitra 2005). Mitra observed that when he placed an Internet-connected computer in a public space, children with no access to formal schooling or knowledge of English would play with the device and eventually learn how to use it. Mitra believes that children are capable of self-organized learning, and his project is an example of Pongratz's thought to make the means or the tools available and create the conditions for people to be curious, playful, and creative.

In the dialogue above, Pongratz is drawing parallels between the Maker/digital fabrication and manufacturing movement and the earlier Internet. Computer hobbyists, entrepreneurs, and small research and development projects in universities initially led the Internet movement. Pongratz highlights the early adopters, the first movers in what he considers a new paradigm for design and fabrication. Before my conversation with Pongratz, I reviewed his Web site at Texas Tech. In some of the course descriptions in the architecture program he developed, I found a concept, "methodologies of becoming," which sounded similar to the idea of "being and becoming," the performance and performatory activity of development (Holzman 2009, pp. 17–20). In the following dialogue, we discuss teaching and methodologies of becoming.

Jim: I don't know how students learn in architecture. Could you describe the challenges and how you go about it?

Christian: The major vehicle that we operate is what we call the "design studio." Everything else, our courses, are supplementary. They just feed areas of interest or knowledge that we traditionally believe to be important to communicate, but the really important thing is the design studio. Typically, three afternoons a week for 4 or 5 hours, you sit with the students and discuss ideas and how to go about a project. It is a process that is very open.

Jim: What's the role of the instructor?

Christian: As an instructor, you focus on the peculiarity of each person. With 15 people, it's 15 different processes and ideas, and 15 various projects. The weird thing about our profession, and that's probably a big drawback, is that it's never been streamlined. It's not like the automobile industry, where, because of individual performance studies and experiments, you know that you have to have a particular type of part in the process. Architecture is not like this. You get a product that is non-engineered. It's personal, customized, individual, from the client side and the designer side.

Jim: What do you mean by "client side"?

Christian: You always have to respect, of course, the interest and motivation of the client. That plays into the game, and you have to integrate what the client wants into your process. Then, when we talk about "becoming," it is also for us, I think, a description of the actual process, once it goes into the computation. You set parameters in the beginning, and then you inscribe a process computationally. In a computational process, you can go back to the beginning and change the initial parameters and run the process again—the idea and the original form, everything about form and geometry and shape that is "becoming." Today, everything you throw into a creative process becomes a computational process that turns into, at one

> point, a frozen moment that shows you one image of where it is at that moment in time, and where it still could go. I'm not surc if that is understandable.

This aspect of our interview leads me to realize that there is a significant connection to the performatory theory of human development that I discuss in this book. The design studio is a performance space, a stage where tools can be used to make other things including new tools. When Pongratz works with his students, together they create interdisciplinary zones of proximal development. He makes his expertise available and offers direction and production advice, but he does not tell them what to do. He is focused on the development of skills and practices. The skills and practices span computation, design aesthetics, architectural concepts, fabrication concepts, engineering concepts, and client relationship concepts.

Pongratz refers to the computation methodology for becoming or a process for generating something new. He is talking about the ability of computer-aided design software to model the effects of changes in the design. With computation, the starting parameters of a model can be set and changed repeatedly. How the modeling software runs depends on starting parameters, and those parameters might change as the design changes, and that might ultimately result in a different object. The process of becoming that Pongratz describes sounds like a theory of being and becoming to describe a similar process for human development. However, there is a problem with the comparison that I've drawn.

Being and becoming involves a process of deconstruction and reconstruction. Playwright Dan Friedman offers an example: "If you are a playwright, the deconstruction process involves questioning your own assumptions, experience, concept of the beautiful and/or interesting; taking them apart and putting them together in the process of writing the script." (Friedman 1999, p. 187). The creative script writing process that Friedman describes seems similar to the process of resetting parameters and rerunning a computational model that Pongratz describes. Parameters might be similar to assumptions and experiences. Re-examining assumptions might be like resetting parameters. But there is a crucial difference. The computational methodology is a linear process, and changes in parameters are generated in logical relationships to a starting point or a reset point. The deconstruction/reconstruction process that human beings are capable of is not linear; it is creative.

A human developmental process of being and becoming cannot be parameterized (reduced to input variables) or systematized (made linear or predictable).

Pongratz identifies the significance of the contributions of architecture clients and the relationships that must be built to include clients in the design process. He describes a collective process where individual development and growth is connected to the materials, place, and people. The instructor's role is to create processes of becoming. This process includes using the different ways that students are developing or becoming as they are designing and suggesting that certain directions may be productive.

Jim: How do you understand the design methodology?

Christian: I think Guattari at one point wrote a paper about the abstract machine (Deleuze, Guattari & Stivale, 1984). Guattari makes the process into a metaphor of methodology that is infinite. It doesn't have an end. During that process, you can throw a lot of things onto it that can be mapped onto a human process. So, you have several creative people around working on something, or you have a machine that uses ingredients to produce something. The only way to understand it is to make snapshots, at instances, to see where that process is. The problem he describes is that we don't yet have a methodology to decide when to stop. When is the moment to freeze perfectly? That's very difficult for us.

Pongratz raises an interesting problem related to continuous design process and applicable to the continuous developmental learning process. In education, we have tried to develop methodologies to decide when to "stop." We call it an assessment, or testing. For example, a learning standard is established that describes what a child at age seven should be able to do. The day the child is tested creates the "frozen moment." The results from that point in time are used to make judgments about what the child knows and whether or not she is ready to move on to the next level. The moment to freeze becomes problematic; the judgments about learning based on a frozen moment transform the relationship to learning. It makes learning systematic. Learning becomes a process in which the difference between what a child knows and what the test measures using the standard as a benchmark is used to make decisions about teaching and learning. The problematic aspect of systematic education is that human beings are always changing and changing unpredictably. There are no frozen moments, and 7-year-olds all develop at different rates. The stop and test approach to research-based standardized assessment oversimplifies the reality of each student and does not provide sufficient information for an instructor to make timely decisions about how to help children learn. In my view, there is no frozen moment in design or learning, and there is no need for one. Our conversation continued into the area of curiosity and failure.

Jim: If what you're saying can be realized, is being in a type of design studio all that is needed for learning from mistakes?

Christian: You need enough curiosity to understand that moment when you fail and automatically move on. What comes to mind right now is the question of failure. In a high school, everything is about needing to have the best grades. The experience that you have if something goes terribly wrong is a key moment. That's where you learn. We don't teach that. We don't want anyone to fail.

Jim: How do we address this?

Christian: That brings us back to playfulness. Playfulness is not even allowed to be part of learning because you have to follow the rules to reach the goal quickly. Instead, if you sit at a table and I give you five ingredients, and I want you to think critically, it's just about making something. Does it work? No, but you will probably go back and start over again if it is playful.

Jim: What happens when people say, "but it won't work, because…." How does that go for you?

Christian: I have fun with that because a lot of times, I don't even understand what they are talking about. I might not have a clue what the science behind something is. When I approach the problem, I look at it through a different lens. I have a more intuitive approach to thinking about the problem because I have a larger goal in mind. I also see a different kind of collaboration in the Maker culture. When a person cannot go further, and someone next to them helps with that step, they just explain it or show it. I think that's interesting.

Our conversation turned to a discussion of "design thinking." It seems to be the new category for approaches that incorporate design, collaboration, creativity, and build activities into work and learning processes.

Jim: Could you say a little more about design thinking? I'm not familiar with it.

Christian: It brings us back to the design studio. You have an idea for a project, and the way you go about it is to look at it from many angles, bringing in clients and materials and environments. You create the first concept to describe the idea. It could be a two-dimensional diagram. It could be a photo—something that can latch onto your imagination. It becomes the driving engine of the whole project. From there, it goes through a process where you make several proposals or prototypes, and you question each one of them. How does it perform against the criteria? You change it, optimize it, and go back. You create alternatives, and the idea goes through alterations. You consider economic, budget, and design questions. This whole process, developed to go from the initial spark of an idea to the final product, is used in a lot in other disciplines right now, and they describe it as design thinking.

Jim: What's an example of an organization that uses this approach?

Christian: A company called IDEO™ is a creative design company that maps into many sectors. They talk a lot about the creative design process, per se. Design is a social process, and it empowers people. The process should contribute and respect the wishes or requests of people. Design thinking has role-playing, scenario playing, brainstorming, "what if" scenarios that allow you to map out potentials. That's what I understand about design thinking.

Design thinking is embedded in STEM learning curricula such as Engineering Is Elementary™, which elementary school teachers are using, and Engineering by Design™, which is intended for high school. Engineering education learning units are implemented in science education as part of a project-based learning methodology in STEM education (Johns and Mentzer 2016; Lottero-Perdue et al. 2016). Design thinking as a design studio pedagogy, particularly scenario use, fosters creative thinking in students (Casakin et al. 2016).

During our conversation, Pongratz led me to believe that what he is proposing is more than using design thinking as a teaching method. He suggests it should be an organizing principle that will open up organizations and school curricula to creativity and innovation. Stanford University has been a pioneer in using design thinking as an institutional process in their d.school at the Institute of Design at Stanford. Stanford provides resources for K-12 education, higher education, and graduate-level certificates in design. Pongratz believes schools organized with design thinking principles would be more receptive to students, would stimulate them and disrupt the ways everyone operates. It would help students and faculty look beyond boundaries and integrate them into a creative process.

I agree with Pongratz, and I see parallels between the design thinking process and the performatory approaches to creating developmental STEAM learning environments that I am proposing. Design thinking is explicit about the procedures for generating a creative process and making products or objects. Human development and learning need a performatory orientation that is explicit about being developmental and not product or outcome oriented. A performatory approach to learning does not presume that people are a certain way and think in a certain way. A performatory approach also supports the idea of being open to others, to the ways that they are different, and to the challenge of building creatively with what individuals have to contribute. In contrast, schools and college courses assume students are a certain way and that they have met minimum requirements. They require students to participate in a certain way and make predefined contributions, e.g., correct answers to questions. Performatory approaches to learning expose the contradictions of the traditional approaches to learning we take for granted. The work of educators will be to address these contradictions if they want to create new kinds of learning institutions that can optimize learning.

Science Education Is Changing

The dialogue with Pongratz creates a larger context for looking at science education, particularly if performative approaches, such as those featured in the CESTEMER conference, are of interest. A 2012 review of the literature on design thinking revealed that design thinking is about focusing on solutions (Razzouk and Shute 2012). Traditional approaches to schooling in the STEM disciplines are about teaching students to remember facts and procedures to solve known problems. Designers are trained to "switch modes of thinking" (e.g., from analysis to synthesis) and to address complex problems (p. 343).

In schools, science and engineering education are being brought together so that students can have more hands-on experiences of combining and creating solutions to problems and conducting a scientific inquiry (Johns and Mentzer 2016, p. 13). One study of project-based learning and design thinking done in college-level engineering courses suggested that these approaches improved student retention, satisfaction, diversity, and learning. Significant challenges were also identified, including cost and faculty commitment to design thinking pedagogy (Dym et al. 2005, p. 114).

The Next Generation Science Standards (NGSS) for K-12 education have already incorporated design thinking and engineering concepts (http://www.nextgenscience.org/). These standards were produced in a collaborative effort among the National Research Council, the National Science Teachers Association (NSTA), and the American Association for the Advancement of Science and Achieve. These organizations coordinated with 26 states to implement NGSS.

The stated purpose of the standards is to make college and career readiness a priority across the country. The NGSS is similar to the Common Core State Standards (CCSS). It is an effort to create national science standards. These standards organize science education into three areas: practices, core ideas, and crosscutting concepts. Crosscutting concepts, such as "cause and effect," are made explicit so students can learn to apply the concepts across knowledge domains. The standards provide guidance on how to engage students in activities that provide opportunities to apply crosscutting concepts. The disciplinary core ideas cover content in Physical Sciences, Life Sciences, Earth and Space Science, and Engineering. According to the NGSS.nsta.org Web site as of February 2016, 17 states have adopted NGSS and are working toward state-level implementations.

Pongratz and Holmes describe performatory visions of interdisciplinary learning and science education. However, educational leaders and research institutions re-enforce traditional learning in the education-for-workforce development paradigm by providing standards that contain "desirable" elements of performatory approaches such as collaboration, hands-on activities, and creativity. Newman and Holzman describe this tendency to standardize, or to insist that the process of education must be framed scientifically, as the "epistemic posture." This posture is in the way of the "revolutionary, humane and developmental move our species needs to make" (Newman and

Holzman 1997, p. 7). I believe the epistemic posture in education marginalizes performatory and creative approaches to learning. The epistemic posture also minimizes the impact of collaborative and experiential learning by insisting on testing knowledge acquisition exclusively and effectively separating learning (for the purpose of measuring it) from meaning-making activities.

References

Casakin, H., Van Timmeren, A., & Badke-Schaub, P. (2016). Approaches in design education: The role of patterns and scenarios in the design studio. *Problems of Education in the 21st Century*, 69 (Caliksan, 2012), 6–21.

Deleuze, G., Guattari, F., & Stivale, C. (1984). Concrete rules and abstract machines. *SubStance*, *13*(3/4), 7–19. Retrieved from http://doi.org/10.2307/3684771.

Dym, C. L., Agogino, A. M., Eris, O., Frey, D. D., & Leifer, L. J. (2005). Engineering design thinking, teaching, and learning. *Journal of Engineering Education*, *94*(1), 103–119. Retrieved from doi: 10.1002/j.2168-9830.2005.tb00832.x.

Holzman, L. (1999). Life as performance. In L. Holzman (Ed.), *Performing psychology: A postmodern culture of the mind*. New York, London: Routledge.

Holzman, L. (2009). *Vygotsky at work and play*. New York, NY: Routledge.

Johns, G., & Mentzer, N. (2016). STEM integration. *Technology and Engineering Teacher*, (November), 13–18.

Ledford, B. Y. H. (2015, September 17). Team science. *Nature*, *525*, 308–311.

Lottero-Perdue, P., Bowditch, M., Kagan, M., Robinson-cheek, L., & Webb, T. (2016). An engineering design process for trying (again) to engineer an egg package. *Science and Children*, (November), 70–78.

Newman, F., & Holzman, L. (1997). *The End of knowing: A new developmental way of learning*. New York, NY: Routledge.

Porter, A. L., & Rafols, I. (2009). Is science becoming more interdisciplinary? Measuring and mapping six research fields over time. *Scientometrics*, *81*(3), 719–745. Retrieved from http://doi.org/10.1007/s11192-008-2197-2.

Perbellini, M., & Pongratz, C. (2015). *Digital media for design*. CA: Cognella Academic Publishing.

Razzouk, R., & Shute, V. (2012). What is design thinking and why is it important? *Review of Educational Research*, *82*(3), 330–348. Retrieved from http://doi.org/10.3102/0034654312457429.

Rylance, R. (2015, September 17). Grant giving: Global funders to focus on interdisciplinarity. *Nature*, *525*, 313–315. Retrieved from http://doi.org/10.1038/525313a.

Sugata Mitra, (2005). Self organising systems for mass computer literacy: Findings from the 'hole in the wall' experiments. *International Journal of Development Issues*, *4*(1), 71–81. doi: 10.1108/eb045849.

CHAPTER 4

Technology

The second, full definition for the word *technology* offered by The Merriam-Webster online dictionary is "a manner of accomplishing a task especially using technical processes, methods, or knowledge." My interpretation of the definition leads me to include language, hand tools, text, non-textual/visual representations, fire, and using a rock to break open a coconut, as examples of technology. In my opinion, any contemplation of technology is also a contemplation of human culture. And our uses of technology certainly span our understandings of the STEAM disciplines.

In the field of instructional technology or educational technology, the use of computers, information and communications networks, multimedia, Internet Web sites, and a vast variety of mobile devices and interactive presentation tools are what is meant by technology use in classrooms. There are also assistive technologies that are specifically used to support learners with a variety of documented disabilities with additional help in the classroom. These might include hearing aids, classroom wide audio systems, text to speech software on computers, visual magnification features for screen viewing, and many other technologies. Whether the definition of technology is specific or broad, what is important is the recognition that human beings use tools to create the things that they need and want. Using tools and creating things transforms us and is the "tool and the result" of human history in a Vygotskian sense (Vygotsky 1978, p. 65).

J.E. Martinez, *The Search for Method in STEAM Education*,
Palgrave Studies In Play, Performance, Learning, and Development,
DOI 10.1007/978-3-319-55822-6_4

Language, "Tool and Result" Technology

In his book, *Orality and Literacy, The Technologizing of the Word* (1982), Walter Ong introduces the reader to primarily oral cultures and describes the transition of human beings from being primarily oral in communications to using text and developing literacy. This transition had a significant cognitive impact; it produced the ability to think abstractly. Previously, human beings focused on remembering (individually and collectively) everything that was known. According to Ong, "[p]ersons who have interiorized writing not only write but also speak literately, which is to say that they organize, to varying degrees, even their oral expression in thought patterns and verbal patterns that they would not know of unless they could write" (Ong 1982, p. 55). What Ong describes can be understood as a dialectical relationship between writing and thinking; abstract thinking emerges in relationship to writing. Ong traces the human capacity to do science, create categories, do logic, and think abstractly to the social transition of people living in primarily oral cultures to literate cultures.

Support for some of Ong's conclusions comes from Soviet psychology, specifically the work of A.R. Luria's *Cognitive Development: Its Cultural and Social Foundations* (1976) and Michael Cole and Sylvia Scribner's cross-cultural investigations of learning in formal and informal settings (Scribner and Cole 1973). As an aside, Lois Holzman was introduced to Vygotsky during her time in Cole's research laboratory, the Laboratory of Comparative Human Cognition, operated at Rockefeller University in New York City. Holzman ultimately went on to partner with Stanford-trained philosopher Fred Newman, and together they developed social therapeutics and contributed to furthering dialectical understandings of human tool use and human development (Newman and Holzman 2014; Holzman and Mendez 2003). Ong seems to share foundational ideas with Vygotskians that describe how tool use changes how we think. Despite some obvious unintended consequences and misuses of technology, in its broadest sense, technology is not bad for us. Technology shapes our thoughts and the social, cultural, and emotional aspects of how we live, and we shape technology.

(a) Historical Accounts—Digital Natives

I dislike the term "digital natives," and I dislike the term "digital immigrants" even more. Marc Prensky popularized the term in a 2001 report titled, *Digital Natives, Digital Immigrants*. In the report, he specified that digital natives are those who were born into the digital age. Those of us born before the digital age are digital immigrants. According to Prensky, "As Digital Immigrants learn—like all immigrants, some better than others—to adapt to their environment, they always retain, to some degree, their 'accent,' that is, their foot in the past." In addition to emphasizing a generational divide, Prensky's report defines a problem, "**our Digital Immigrant instructors, who speak an outdated language (that of the pre-digital age), are struggling to teach a population that speaks an entirely new language** [his emphasis]" (p. 2). In his report, Prensky provides numerous anecdotes of how digital immigrant teachers are failing to reach digital native students because of their "outdated" language. Prensky is partially correct in identifying that there is a problem. However, the problem is not rooted in a pre-digital/post-digital divide or a generation gap. What I dislike most about Prensky's rhetoric is that it works to separate students and teachers. The so-called digital natives are disconnected, and my generation (people born in the 1950s and 1960s) is given a misleading label when, in fact, my generation did the work that gave rise to the digital age.

Prensky's ideas are ahistorical and distracting. According to him, the "digital age" and "digital natives" are somehow separate from the pre-digital history from which they emerged. The assertion that the significant difference between one generation and another is fluency in the new culture is somewhat dismissive of the complexity of the experience of immigrants (of all kinds). It is also a simplified representation of how cultures educate and produce technical fluency from one generation to another. I am not alone in my critique of Prensky. For example, in the March 3, 2010, issue of *The Economist*, the utility of Prensky's ideas is challenged for their oversimplification in describing an entire generation and the complex social interactions in educational institutions. Despite Prensky's simplistic explanation of what goes on in schools, many teachers and teacher educators have latched on to the "digital natives" idea. I believe there is an alternative to creating generational divides. Educators could work together with students to build understandings and ways of relating that would not be possible without the unique histories that each generation brings to the learning setting.

Andrea's Tech Crew

The Young Women's Leadership School of Astoria (TYWLS) is an all-girls school for grades 6–12 located in Astoria, Queens in New York City. This school is a member of The Young Women's Leadership Network and the home to an innovative STEAM education program led by Andrea Chaves. Chaves is a certified Spanish language teacher with an additional certification in educational technology. She also immigrated to the USA as a young woman from Columbia. Her innovative efforts in STEAM education were recognized in 2016 when she received the White House Champion of Change in Computer Science honor. In addition to the Champion of Change honor, Chaves was also recognized in May 2015 by the National Center for Women and Information Technology (NCWIT) with an Aspirations in Computing Educator Award. She also received the Empire State Excellence in Teaching Award from the governor of New York State.

Her teaching position with TYWLS network of schools affords her access to corporate partners and a successful model of educating girls and young women, which has created opportunities for Chaves to innovate within and outside of the classroom. In the interest of full disclosure, Chaves also happens to be a former student of mine.

In 2012, Chaves began working with girls in the middle school and high school to create a Tech Crew program. The Tech Crew provides technical support to the school's instructional technology, the annual Digital Dance program, an annual Eco-Friendly Fashion show, and a summer program called TYWLS Tech Explorers. Tech Explorers is a 3-week-long tech camp that provides opportunities to learn how to program computers. Chaves currently works with up to 90 students each year. Her Tech Crew program is so popular that there is a waiting list of students who would like to participate. Tech Crew students employ a mentor/mentee model during a 2-week period in December called Intensives. During this time, class schedules are suspended, and teachers and students create new kinds of learning experiences. Tech Crew members work with other students to create a program called Digital Dance. The Digital Dance broadens participation in technology and computer science learning by integrating the arts (dance, multimedia, set, and costume design) in a public event that is attended by the community. The Eco-Friendly Fashion show focuses on promoting the importance of recycling found items and environmental awareness. The fashion show integrates technology, and the Tech Crew gets involved in producing the technology aspects of the show. The fashion show includes the entire school community. The Tech Crew students have also used the popular computer programming education Web site Code.org as one of their tools to organize the annual Hour of Code, which opens the Intensives period in the school and engages all 562 students in an hour of learning to write computer programs. Chaves started out by working with organizations such as NCWIT and Code.org. She recognized that the Code.org program, as good as it was, could not sufficiently address the particular circumstances and interests of her students, so a "build it ourselves" strategy was adopted. The plan included the students as builders of the formal and informal educational programs. According to Chaves, one of the fascinating aspects of this model is that younger girls take responsibility for mentoring older girls. Collaboration across grade-levels challenges current concepts of age-based grouping in public school environments. Chaves asserts that her students have been instrumental in program innovations, and they have been significant partners in her in day-to-day problem-solving efforts. Many of her former students are now in college and are seeking opportunities to give back and expand Tech Crew's impact on learning. Chaves has documented evidence of successful learning outcomes and students graduating high school and going on to college with scholarships in the STEM majors. Four other schools in the Young Women's Leadership Network are interested in using the Tech Crew model.

Chaves describes a trial-and-error process for working with students and external partners. While her students have been enthusiastic builders of the programs they participate in, she has been unable get other teachers to commit time and expertise consistently and reliably. Chaves provides us with an example of what is possible when we do away with the artificial distinctions that separate teachers from students. The opportunity that Chaves seized upon to create her programs with students was the two-week-long Intensives program that is a unique feature of schools in The Young Women's Leadership Network.

Chaves takes her program design cues from the external partners. For example, she observed that one partner had internal project managers assigned to different projects running small groups. She incorporated that concept in Tech Crew. Another prominent feature of Tech Crew is the mutual respect that arises from the mentoring model where girls are responsible for teaching each other. Chaves believes that the culture of Tech Crew produces strong relationships that result in students becoming passionate about learning. Chaves has no formal training in computer science or small business, yet she received a Champion of Change honor for creating developmental learning environments where young women learn to write computer code and provide technical support to their school by learning through creating new performances.

Chaves attributes her success to letting go of power in the classroom and giving the power to the students. She takes on what she describes as an unobtrusive supervisory role. She cannot point to a definite influence in her personal history that suggested to her that distributing leadership to students was a viable strategy. Chaves is an ordinary teacher who is doing extraordinary things with her students. She is innovative and creative and willing to risk making mistakes. From my observations of her with students, I can tell that she loves her students and that her students love her. She has created a performatory environment and a culture where all kinds of zones of proximal development are possible. The Tech Crew's activity is creative and performatory and has fostered learning and development for Chaves and her students.

Jose Santiago's Family

Jose Santiago has an undergraduate degree in psychology. His first job out of college was as an attendance teacher in the New York City Board of Education. Shortly after the reorganization of the New York City Board of Education into the Department of Education, Santiago moved out of New York, took a job in a healthcare corporation in Florida, and diversified his skill set. He worked as a technology support specialist for about 9 years, managing a computer network of over a 1000 users. After relocating back to New York and working for another healthcare corporation, Santiago decided to pursue the teaching certifications that would enable him to return to the classroom. He currently works in a Brooklyn public school with approximately 500 middle school students. Santiago is the official NYC Department of Education Single Point of Contact (SPOC) for technology support at his school.

According to Santiago, the Tech Team at his school started organically, with one eager and exceptional student and a schoolwide need.

Santiago explained that the student would hang around the school Tech Lab during lunch breaks. Santiago allowed the student to accompany him on visits to classrooms to complete minor technology repairs. Santiago had more equipment around the school than he could maintain, and the student was eager to assist. The student proved to be exceptional at completing the minor repair jobs he was assigned. The student created trusting relationships with the teachers in his school. Teachers were impressed with this student, and Santiago was desperate, so he formalized the student's role. That student's success was transformative and created opportunities for other enthusiastic students to volunteer to do technical support for the school. Santiago's first Tech Team was composed of five students. Today, 14 students, who complete a rigorous selection process, currently staff the Tech Team. Demand to participate in Tech Team was so high that a classroom-based embedded team model is used to provide more opportunities for students to provide in-classroom technical support services. These students support routine technology use, e.g., removing and replacing laptops from the laptop carts as part of the classroom routine. Approximately 60 students serve on these small classroom-based teams. Santiago offers the following explanation of why he thinks students are so eager to be involved in Tech Team-related activities.

> They love technology. Every kid loves an iPad or computer; put it all together and a family is created. It's amazing. The tech students are drawn from the 6th, 7th, and 8th grade. As kids step up, they train the younger ones. I have them in the morning as a homeroom class. In the morning, they greet each other like family, and when they are together, they are a tight unit. The students come from different places and socio-economic backgrounds. They get a feeling of routine and normalcy, and this is more than teaching technology skills and knowledge.

Santiago explains that the students created the "family environment" concept, and he bought into it. Santiago cannot identify any one factor that made it possible for him to introduce Tech Team innovation into his school. He says that the initial success with his first group translated to formal support from school administrators. The additional effort required to organize the students into a Tech Team is also a voluntary contribution on his part.

Santiago encourages students to think on their own and to use information and conduct inquiries on fixing problems. Santiago believes that there is carryover to the academic subjects; students understand that they are representing the team, and they hold each other accountable. Santiago has created a trusting environment. Students know that Santiago has their best interests at heart and that he will intercede in difficult situations. Santiago says, "We need to teach kids how to advocate for themselves, give them a voice, responsibility, and demand that they work hard and be honest."

Chaves's and Santiago's successes are rooted in collaborations with students. They presented students with real-world problems—problems at school that required solutions and needed people to solve them. Chaves and Santiago are technology teachers in their schools. Having been a public middle school technology teacher myself, I can offer some insight into the factors that I believe contributed to their success.

Technology teachers are responsible for the maintenance of nearly all of the equipment in their schools. Tech teachers often receive additional hours of release time from regular teaching duties each week. Tech teachers use lunch hours and contractual preparation periods to maintain equipment in their laboratory spaces and to visit the classrooms of other teachers to fix problems is a common strategy. Tech teachers are highly visible in the school community, and they have flexibility in their schedules that regular classroom teachers do not have. Students are interested in Tech teachers for the access to technology that they can provide. Tech teachers usually create their class projects and curriculum, and they decide whether or not testing is required. They often team up with other classroom teachers to support the integration of technology into classroom projects. In short, when good Tech teachers are involved, students expect fun, collaborative projects, and access to technology. Tech teachers usually become the popular teachers in the school.

Time, flexibility, opportunity, administrative support, student involvement, and a willingness to take risks are all significant factors in contributing to the innovations that Tech teachers can bring to a school. However, it was the willingness of Chaves and Santiago to collaborate with students across the digital and generational divide that was transformative. They created new performances for themselves and the students.

References

Holzman, L., & Mendez, R. (2003). *Psychological investigations: A clinician's guide to social therapy*. New York: Brunner-Routledge.

Newman, F., & Holzman, L. (2014). *Lev Vygotsky: Revolutionary scientist: Classic edition*. United Kingdom: Psychology Press.

Ong, W. J. (1982). *Orality and literacy: The technologizing of the word*. London: Methuen.

Prensky, M. (2001). Digital natives, digital immigrants. *On the Horizon* (Vol. 9). Retrieved from http://www.emeraldinsight.com/10.1108/10748120110424816.

Scribner, S., & Cole, M. (1973). Cognitive consequences of formal and informal education. *Science*, *182*(4112), 553–559.

Vygotsky, L., Cole, M., John-Steiner, V., Scribner, S., & Soberman, E. (1978). *Mind in society: The development of higher psychological processes*. In M. Cole, V. John-Steiner, S. Scribner, & E. Soberman (Eds.). Cambridge, MA: Harvard University Press.

CHAPTER 5

Engineering

What's This Maker Stuff in the Schools All About?

A colleague of mine asked that question. She's a career educator with a background in the humanities. I opened my mouth, and words stumbled out, but they did not make any sense to me until I was able to say, "It's like shop class but for profit." My colleague and I are both old enough to remember middle school wood shop or metal shop classes. These classes were "specials," like gym, cooking class, choir, and band practice. Students got to choose these classes based on interest. Wood and metal shop had basic hand tools and powered equipment. Today's "Maker spaces" are the twenty-first-century versions of a metal shop class that includes everything from computer-controlled 3D printers (devices that extrude melted materials in three dimensions) and wood routers to soldering irons and sewing machines. A popular idea in schools is to provide hands-on Maker space activities that are fun, promote skill development, are educational, and can be done without powered equipment. In some school districts, library spaces are being converted to Maker spaces.

Maker spaces outside of schools use a club membership model for funding rental work and storage spaces for Makers to use and share equipment. The equipment is available to be "checked out," like library books. Members pay for or provide materials for fabrication purposes. The culture of Maker spaces features collaborations that encourage learning by doing. Makers share expertise and create partnerships with those who have complementary skill sets to develop ambitious projects. Maker

J.E. Martinez, *The Search for Method in STEAM Education*,
Palgrave Studies In Play, Performance, Learning, and Development,
DOI 10.1007/978-3-319-55822-6_5

culture is often described as a subculture and compared to hobbyist and computer hacker cultures. The idea in a Maker space is to learn to use a variety of fabrication tools and materials to make things. There is an entrepreneurial attitude behind the Maker movement, if a person can learn to make things and fabricate products, those products might have commercial value. I think that schools will attempt to adopt Maker activities without committing to the attitudes and entrepreneurial culture that gave rise to the Maker movement. I view making in the Maker movement as developmental meaning-making. Unfortunately, I can imagine that many will apply "stop and test for content knowledge" model of learning to Maker activities. My hope is that assessment strategies might focus on "what students have learned" as part of ongoing integrated project-based Maker activities in classrooms.

The Maker Movement and Engineering in Education

The Maker movement is composed of engineering enthusiasts, artists, craftspeople, and entrepreneurs who use small-scale fabrication tools, such as 3D printers and microcontroller technologies, to create small businesses. It is a "do-it-yourself" entrepreneurial environment that is finding its way into formal and informal education settings. The most prominent Makers are entrepreneurs who are writing books that sell their products as well as educate the public. I have observed that the Maker/do-it-yourself for-profit movement seems to be targeting middle-class students as the consumers of their products and educational materials. Many educators and informal education providers view the Maker movement as an entry point to an engineering education. My interest is in providing the broadest grouping of students with access to the latest innovations in education. My primary concern is on how to make these ideas, tools, and practices accessible to all of the students who might not be able to afford a Maker club membership or a well-equipped workshop in a basement or garage.

Traditional Path in Engineering

As a professional discipline, engineering has many specialty areas too numerous to list but as diverse as electrical engineering, chemical engineering, civil engineering, bioengineering, environmental engineering, and genetic engineering. Engineering has become interdisciplinary as a professional practice. The preparation of engineers begins in high school. Students with strong academic standing in mathematics, physics, and science apply in a competitive application process to schools of engineering. The preference is for students to gain acceptance to an ABET Accreditation Board for Engineering (ABET) (Accreditation Board for Engineering and Technology) engineering program. Once they are in an engineering program, students take advanced courses in mathematics, physics, and applied sciences that are unique to the engineering specialty of interest. Upon graduation, entry-level jobs are obtainable. However, obtaining professional licensure is recommended. Professional certification requires taking an initial licensing examination and time spent in internships with professional supervision.

Some students who exhibit early interest in engineering have opportunities in middle school and high school to participate in specialized programs. Often these middle schools and high schools have competitive application processes, and students with high scores on standardized academic achievement tests in mathematics and science are successful in gaining entry to specialized schools. The underlying assumptions in the engineering career pathway are (1) engineers are expected to be competitive, (2) engineers are good at math and Science, and (3) engineers self-select into competitive engineering programs as high school seniors. While there is a definite movement, backed by government grant funding to increase the numbers of young women and minorities entering the engineering fields, there may be some work yet to be done on developing alternative pathways into engineering.

Informal Conversations with Two Young Women

In 2012, I co-taught a Career Discovery course in the School of Engineering and Computing Sciences at NYIT. The course was for freshmen and had a service-learning component. There were about 24 students in the class, including five young women. I was managing the service-learning aspect of the course and would occasionally travel with students to the service-learning site. On one occasion, I went with a young woman who was planning on declaring computer engineering as a major. While making small talk on the train, I asked her about her interest. She thought computer science was fascinating and felt it might be a good career. I asked her about her experiences with computers in high school. She said she used computers, but had never opened one up, and she was just learning to write programs. A couple of weeks later, I had another occasion for small talk with another young woman, and she had a similar story: She was counting on studying engineering in college to get a job. When I pressed her on why she wanted an engineering career, she also said that she was interested in the field. I compared their stories to those of the young men that I was getting to know. Many had already been encouraged since childhood toward the engineering field by parents and teachers. Many were already accomplished programmers, and some had been building computers and repairing mobile devices for years. The young men volunteered information about how parents pushed them into engineering, or how they just felt it was the right field because they liked fixing things or because they saw computer software development as the means to Internet start-up riches. Young women and young men were coming/seem to come to engineering with different stories and interests.

My accounts are not representative of the many different stories of young women who become engineers. The young women presented in my stories qualified to be in an engineering school. They satisfied the entrance requirements in math and science achievement and SAT scores. Engineering is a competitive professional environment, and engineering school is competitive as well. It seemed to me that these young women who were pursuing computer engineering degrees would have to work hard to catch up on the hands-on experiences with hardware and software that their male peers had. In what follows are some developmental approaches that innovative educators are taking to creating opportunities for students to engage in hands-on STEM and design activities in primary school settings that might help them compete later on. While these educators may not use words like "performance" what they describe are new performances in school for themselves and the children they work with.

The Magnet School Approach

According to the U.S. Department of Education Web site, the Magnet Schools Assistance program, which provides funding to public schools, is designed to "reduce, eliminate, or prevent minority group isolation, produce high student academic achievement, promote diversity and socio-economic desegregation and deliver innovative theme-based curriculum" (U.S. Dept. of Education 2016). One of the themes that Magnet School administrators have prioritized is the creation of STEM-themed schools. There are significant challenges for educators who take on turning existing schools into STEM-themed schools. These challenges include organizing teacher professional development, integrating the new theme with existing curricula, identifying and working with external partners, sustaining related practices, integrating new technologies, improving parental and community involvement, developing new marketing materials for attracting diverse families to the school, and addressing issues related to improving student academic achievement.

Ellen Darensbourg is an educator who has been the STEM Magnet specialist at two different elementary schools over the last 6 years. She has been an educator for 24 years. Her undergraduate degree is from Brandeis University in Boston, and she completed her graduate teacher preparation at Bank Street College in New York City. She was part of the second cohort of Teach For America teachers, and her first teaching assignment was in Los Angeles, in Compton in 1991. In our interview, she provides accounts of her efforts at creating elementary schools that support children to develop interests in engineering in sustainable ways. Preparing teachers to take on specific themes in Magnet schools takes time and effort. In the dialogue that follows (edited for clarity), Darensbourg shares some history of her preparation for school leadership and becoming a Magnet Specialist.

Jim: What do you think are the barriers in preparing teachers for STEM education?

Ellen: Fear. I think a lot of teachers are fearful of those fields, either because of a previous bad experience or lack of confidence. One of the first things I hear teachers say is "I don't know anything about science," or "I never taught science," or "I'm not good at math."

Jim: What kinds of STEM learning experiences are important to student learning?

Ellen: I think the informal and relaxed learning experiences are more powerful, where students don't even know they are learning. They are not worrying about memorizing something. They're just doing. For teachers and students, the power of STEM is the hands-on, "just get in there and do it" attitude. The whole theory behind Engineering Design Process is just going for it. Go ahead and try it out and don't be afraid to fail because that's part of the process. The idea that you have to learn something and be good at it just goes out the window because you just have to do it.

Jim: Engineering Design Process is a particular thing you're referencing?

Ellen: Yes. Engineering Design Process is the process through which people solve problems—first and foremost, through engineering. But it is a process that applies across the board. It has ties to a writing process, a scientific process. It's how people approach problems. You go through it, you think through it, ask questions about it, do a little research, and try it out. You look at different possibilities. Create something. If something doesn't work, you go back and improve it. And at some point, you celebrate and share your findings.

Jim: Where did you learn about Engineering Design Process?

Ellen: I first heard of it when we were introduced years ago to Engineering is Elementary, an engineering program for elementary schools. We used a lot of their curriculum, and they teach Engineering Design Process. It just became a mantra for us. In my current school, the principal believed that we should own the design process, and we should internalize it and use it as a guide for everything we do. I agree, internalizing the design process is internalizing a method for success in life.

Jim: I want to understand how you figured out how to do STEM teaching. Why are you not one of these teachers who is afraid of learning something new?

Ellen: It started back in 2009 or 2010 when my school in Harlem was asked to participate in a Magnet grant. My principal at the time had no idea where to go for picking a theme, and someone suggested STEM. At the time, I was focused on literacy, and she came to me and asked me to do some research and report back. I just dove in and ended up getting excited about it, but I also discovered that it was a very new area and there was nothing regarding elementary school and STEM. Nobody was examining collaborative STEM learning in elementary school.

Jim: You got released from classroom teaching to do that?

Ellen: I was in already in a coaching position and was doing a reading specialist job in the mornings and data specialist in the afternoon. I wrote a portion of our grant for the Magnet school. To write that grant, I had to do a lot of research.

Jim: You haven't mentioned this explicitly, but could you talk about leadership? What's your relationship with that?

Ellen: When I was learning how to teach, I looked to teacher leaders to help me. I realized the value of other teachers. I made the decision to do the same for others. That's success in education, teachers working together and supporting each other. I've certainly had opportunities to become an administrator, and I've always resisted that role because I felt like the closer you are to the people you work with, the better you can help them. So, being a teacher leader instead of an administrator was more appealing to me.

Jim: You are referring to the distance between teachers and administrators?

Ellen: I discovered that after I had been out of the classroom for many years. Before getting the Magnet grant, I was reassigned on short notice to a classroom teacher position when a teacher quit the first day of school. I was shocked and not prepared, but it was the best thing that ever happened to me. At that point, I had been out of the classroom for seven years. I had been telling teachers what they could and couldn't do, but what I knew didn't necessarily have the same meaning anymore because the expectations for teachers had changed, and the amount of work and data collection changed. Going back to the classroom and experiencing that firsthand was mind-blowing. I think it's important to stay close to what you do.

Jim: Going back to barriers for young people having STEM careers, what's your take on what the obstacles are?

Ellen: I think the biggest barriers would be knowledge and awareness. In pushing STEM education in elementary schools, I made it a mission to talk about careers. Not just to teach the content, but to make connections to the real world and the jobs in those areas. It's important for children to start hearing what a marine biologist is or what an engineer does. We have to put that on their radar. We also have to work at opening the doors to help them. Financial difficulties also come up way too often, and students need the financial support.

Jim: What are politics as they relate to schooling?

Ellen: It is my belief that public education is the key to everything. My mother was a public school teacher. She believed in the system and made sure that we went to good public schools. I've taught in underserved areas my entire career because I think that's where the children are who need opportunities. I guess my politics are that public education needs to be the best it can be and our kids need access to everything that can help them and support them.

Jim: Could elaborate on fear of change in schools?

Ellen: The schools I'm in need change. Even when the principal is not interested in change, I've been able to get in there and say, "I'll do it, I'll make you look good." Schools always need to look for ways to change. If you're only talking about literacy and math, you're looking at one little box. There are so many other ways that we can teach our children and open their eyes to the world around them. Nobody gets anywhere in life just because they had high English language arts and math scores. That's not what makes a person a person. I think that's one of the reasons I fell in love with STEM. It broke my focus out of the box. I still think reading, writing, and literacy are important, but we have to learn to explore life to provide a well-rounded education.

Jim: How do you measure success in STEM education?

Ellen: I would guess how I measure success in the small details as well as in the big ideas. Something as simple as seeing children engaged, excited, involved, and asking questions. After many years of not seeing that in children, then seeing what we do with them with STEM is significant. It's about having kids talking about what they know, sharing, wanting to show off what they've done.

Jim: How are parents involved?

Ellen: Students tell their parents about STEM projects and get them to come to celebrations or family night. Those parents come and say, "It's snowing, but I had to come because my child said I had to and I want to." Parents see the change in their children and are excited. I hear teachers say they're excited again. Seeing the passion in their eyes is wonderful. I see that the children have grown at the end of the school year, not just academically, but emotionally and socially. At the Harlem school, I never felt like STEM caught on, and now I'm working in a school where I feel it has. I used to twist arms to get teachers to come in during vacations time for professional development. This year they want to do it. They're willing to do it. Does it have to do with the fact that they'll get paid? Sure, but I also think this comes from a different place. Now they're willing to stay after school, come in early, spend lunch period to prep. Just seeing the results and their transformations as teachers, having more teachers sign up for professional development on Saturdays, to me that is a success.

Jim: Let's go back to the Harlem school. Was there any aspect of that experience that was a success?

Ellen: Personally, yes. It propelled me in my vision, my belief, and my desire. A lot of what I created there, in conjunction with others like you or on my own, was written into the new grant that I'm in, and I continue to find personal success. There is also a feeling of empowerment that I know what I'm doing, that other people want to hear what I have to say. I can go somewhere else and find a way to create more success. But, it wasn't all about me. Success comes down to leadership. Administrators at the Harlem school were either not interested or didn't realize that they had to commit.

Jim: They didn't realize they needed to commit?

Ellen: It was "Ellen's thing." The Magnet grant is Ellen's thing. The STEM is Ellen's thing. It wasn't Ellen's thing. I was just a voice and a support and a bridge. That's all I wanted to be, and for them to cross over and make it their own. The STEM theme has not continued at that school, and the administrator retired when I left. There were too many problems with the school itself. It had nothing to do with STEM. There needed to be a vision, and there wasn't a vision at all, so the school never really stood a chance. My hope was that STEM would be the breath that gave it more life but, no.

Darensbourg identifies one of the fundamental problems in STEM and other innovative initiatives, sustainability. As a Magnet specialist, she had a leadership role at the Harlem school and was instrumental in getting the STEM theme integrated into the school's practices. However, the administrators never incorporated the STEM theme into the school's overall vision and mission. Others at the school did not invest in the same way that Darensbourg did. As a result, teachers and administrator were satisfied with marginalizing the STEM funded themes and projects as "Ellen's thing." When the funding and Darensbourg left, the school had no resources or champion to continue efforts at changing teacher practices to focus on STEM learning.

Magnet School Details

Jim: That school was 100 children more or less, right? Do you think that was too small of a setting?

Ellen: No, I don't think size matters at all. I believe that it's about values. You either buy in, embrace it, and believe in it, or you don't. There were a few teachers who did believe in it, and the new principal there believes in it, but she doesn't have the training. The change process occurs over time and happens with development and building up knowledge and understanding. If people haven't had that, they aren't going to figure out how to do it on their own.

Jim: I remember that school had dozens of programs going on all at once. Lots of professional development, several partners. NYIT was one. What other programs were in place there?

Ellen: We had Engineering is Elementary. We were writing our curriculum. At that point, we weren't calling it project-based learning. For example, we were expecting teachers to teach Science and social studies when they had not necessarily done that before. It's about the connection and application of content from one area to another. Engineering is the application of science and math. How do you honor STEM without thinking about it that way? There was grant money, and there was an expectation that we were going to write lesson units. How we wrote the units was up to us. With all the research I had done, I decided we had to integrate. We had to make connections across content areas. We had to give purpose to what we were learning and make connections to the real world. Otherwise, we would be wasting our time.

Jim: Can you describe a science and engineering integrated social studies lesson?

Ellen: Second grade has a social studies unit, "New York City, Then and Now." It's essentially their first entrée into history. We have to find a way to make it interesting to a seven-year-old. Second graders are also learning *force* and *motion* as topics in Science. We looked at where there would be a connection between these two things. We came up with the idea of transportation. By learning about the history of transportation in NYC, we can learn a lot about the history of the city. We started with horses and carriages and then switched to wagons and eventually buses, subways, and cars. As our city grew in area, people had to get from place to place. We could also learn about the science behind transportation. Now we're suddenly integrating social studies and science for the purpose of understanding how our city grew to what it is and how it's capable of growing. This also provides students with an opportunity to think about the people and careers involved in making things move and in building cities. Then, we give students some form of a challenge as a culmination of the unit, either to design buildings or transportation vehicles. We give them an opportunity to show that they understand how things move and what would be necessary for our cities to grow.

Jim: How would they show that?

Ellen: Students have created a rollercoaster type transportation system because, for kids, that's fun. The beauty of that system is instead of having to go to the next subway station, the station is outside your window. The kids understand that rollercoasters get people from high to low places efficiently but also make movement fun. They've also created model cities to look at where things might grow, even going so far as to look at actual locations in the city. In an earlier unit, kids looked at the geography of NYC and at the composition of soil and ground and earth to understand what would be good places to build. Where could they put bridges and buildings? Is the land stable enough? We combined all of these things to show them what they could do. They could be engineers and design the technologies.

Jim: When you show them the geography of New York, what are you showing them?

Ellen: Computer maps, Google maps, being able to look back in time and see an area grow. Some helicopters have done video shoots to show different viewpoints. Also, looking at the earth itself. They look at sand, soil, the differences between them, touch them, use magnifying glasses and water filters. They are experiencing things and get to know about geology and geography and not just reading about topics in a book. They become geologists in a way, analyzing land samples or soil samples.

Jim: What happens in third grade?

Ellen: We focus on different aspects but stick to the same theme of the city itself. Students start looking at how we can make things move by using a pulley or a wedge or a lever or an axel. They can create vehicles if they want to. It's open-ended. The social studies piece changes; now they're looking at communities around the world. We have them exploring how people live in different parts of the world. They consider energy, how people use it, waste it, what's renewable and what's not, how we can make recommendations to our community to make our homes greener. We bring in an architect to work with the kids, to talk about designing green homes. They can look at how communities in less developed nations might not even have electricity.

Jim: All of these things sound labor intensive, as far as what the teacher has to do to prepare the lessons. Is that the part that broke or stopped working after you left the Harlem school?

Ellen: This is where your comment on a small school hits on the problem. You have 100 kids in the school, one class in every grade. Each teacher is working in isolation. No matter how collegial the school is, teachers can't help each other much if they are teaching different grade levels. In the Queens school I'm in now, we've got four to five classes per grade and five or six teachers that are supporting each grade. I start off very involved, when I step away, the teachers are there to support and help each other. In fact, this summer, they are working on curriculum. Last summer, I was there 100% every time they met to provide guidance, support, and to give suggestions. Now, I'm not even there.

Jim: Scale helps with sustainability?

Ellen: Yes, it does. It's very unusual to have a school as small as that school was. It happens, and it's possible to succeed, depending on the right people. It makes a huge difference to have a collaboration of teachers who work together. Even if one person isn't as interested or doesn't know as much, there's someone else who can take the lead or somebody who can contribute ideas.

Jim: Let's talk specifically about what is different in the school in Jamaica. Is scale one of the things that makes a difference?

Ellen: Right. The more people you have, the more likely it is you'll have something to work with from the start. So even if it takes longer to get more people involved, you have that possibility from the outset. In the Jamaica school, the administration supported me and jumped on the bandwagon in a much shorter time. The Harlem school was supportive from afar because they were letting me do it. In May, we (teachers Jamaica school) went to a STEM conference in Minneapolis, and the principal came. I think that was an eye-opening experience for her. She dove in. Her excitement level rose, her commitment and engagement increased tenfold. I was able to have a conversation with her about what we could aspire to. This past year, she has been all in. In fact, she came up with our new motto in conversations with me. I kept saying we have to do it; it's who we are. She said, "Right, it's who we are, not just what we do." We were doing it, but we weren't living it.

Jim: What has the impact been?

Ellen: Having administration fully commit had a trickle down effect on the teachers. The administration embraced it wholeheartedly and said that we can teach with Engineering and Design Process as our purpose and framework. But we didn't push it on people, we only suggested it. The school is fully committing now. The reason we're able to do that is that the *teachers saw the difference in the students*(see Chapter Two). They were beginning to see success, and they found it in themselves. They were more excited to teach. Their students were more excited to come to school. Things were falling into place. The learning environment is collaborative now, and there's a lot of shared leadership and empowerment among the teachers. Now, even when I feel I want to assert my leadership, I've found ways to back off and *give teachers more autonomy.* I work as an agent of change, to get things going, but I back off and provide support where necessary. I have to work myself out of a job, which is what a good coach does. I'm okay with that, and I think that's important.

Jim: Can you talk about testing and assessment in STEM education?

Ellen: So that's another aspect that we're looking at this year. How are we sure that we are correctly assessing? You might teach a concept in Science that might not end up in a product. So, how will you assess it? It's a big task.

Jim: What are you assessing, knowledge or some other type of ability?

Ellen: You want to understand how much content the students are getting. But, if we are talking STEM, you should be talking twenty-first-century skills. Students have to communicate and collaborate. How do we assess communication skills? How do we teach those skills? We had kids talking about what they were learning, but we weren't teaching how to present, how to speak, and how to listen to one another. Speaking and listening standards are important. I may not be a fan of all Common Core State Standards, but you can't ignore the fact that those communication standards are there and so prominent. We are working on how we are teaching students to present information in a written form and a spoken form, and how we teach them to listen to one another.

STEM Learning and Special Education

Gina Tesoriero is special education teacher who provides STEM education experiences for students. She is a New York City public school teacher and holds a graduate degree in special education from Hunter College in New York City. Tesoriero is a member of the New York City Common Core Math Fellows and has collaborated with the STEM department of Common Core Fellows to produce The New York City STEM Education Framework. She has received extensive professional development at the New York Hall of Science and Urban Advantage NYC. She is a co-author of two publications and has presented at national conferences on STEM education.

Jim: How did you start doing STEM in schools?

Gina: I've been teaching English, math, social studies and science for the last ten years to students with special needs. Over those ten years, I realized I needed to gain content knowledge in science, so I started to participate in professional development opportunities. I learned a lot about Science through organizations like Urban Advantage NYC (www.urbanadvantagenyc.org). I also worked with the committee that developed the Common Core standards for Math in New York.

Jim: What other training have you had?

Gina: I am certified in special education, and I'm highly qualified in the STEM content areas. I now work for Urban Advantage as a lead teacher, so I get to collaborate with scientists all the time, that gives me a lot of exposure. I received some training at the New York Hall of Science in design thinking in a program called Design Fellows. We had to create projects; mine had to do with biomimicry.

Jim: Biomimicry is?

Gina: It's the study of biology of humans, animals, plants, and natural structures to enhance design. For example, there is a train somewhere in Asia, it was one of the fastest trains ever, but it made too much noise when it went through a tunnel. The engineering team studied a fish that dives in and out of the water without making a sound. They re-modeled the front of the train using the shape of the fish's nose and the redesigned train made less noise, used less energy, and went faster. Taking ideas from nature is engaging for kids. I have a lot of instructional units around biomimicry. I presented at the NSTA (National Science Teachers Association) conference on biomimicry because it's engaging for the students.

Jim: And where did you learn about biomimicry?

Gina: I was just always interested. I worked with the Bronx Zoo and the New York Aquarium, and I wanted to find a way to include design with my work. I've always been interested in animals and engineering, too. I think I'm secretly an engineer and never knew it. I always want kids to have an opportunity to learn about engineering. Traditional approaches to school-based learning are hard for me, but engineering is not.

Jim: How did the biomimicry project work with your students?

Gina: I was teaching in a self-contained 12:1 (12 students, 1 teacher) special education class at the time. We started doing mini design challenges with biomimicry (the students earned them as rewards in my behavior management system). I had a student, let's call him Martin. He had a very low IQ, and he won the first challenge. He eventually won all the challenges. Martin was very creative, and he became a leader in the classroom, though he had never been a leader. But his success also made him aware that he didn't know the same things other students knew. Before learning to design, he didn't know that things could be different for him. When he did realize that learning in school was hard for him, we worked on things that posed difficulties. It was sad to open his eyes to it, but it was also positive because he discovered that he was good at something.

Jim: Wow, that's powerful.

Gina: Yes, Martin was great. I have pictures of him. I think about him a lot. That's why I wanted to do what I do now.

Jim: Are there any other discoveries with special education students to share?

Gina: My students made this thing that was neutrally buoyant. It didn't float or sink. I tried it for a week at a training institute I attended, and I couldn't figure it out. The students were able to create models by studying neutral buoyancy in penguins and lionfish. I couldn't believe they succeeded, where I couldn't. I tried it for a whole week with adults, and we couldn't do it. It was just crazy. Initially, I did the minichallenges as a reward in my behavior management system. When I reflected on how I was using these learning activities, I became frustrated; I wanted to teach in this new way all the time.

Jim: Do you collaborate with other teachers?

Gina: Yes, a general education science teacher. She had an ICT class (Integrated Co-Teaching class to support students with disabilities in a general education classroom) and an English language learner class. We started to talk about collaborating, and we went to our principal. The principal initially said no, but I didn't give up. I don't give up. We prepared some statistics and a curriculum and made our case. Eventually, she said, "this is exactly what I wanted to see." She wanted to offer it to accelerated learners. We were hoping to sneak some English language learners in there because we thought the amount of talking, collaborating, and presenting we would be doing would help English language learners. We were allowed to bring girls into STEM. Seventy percent of those girls are now in engineering-based high schools or programs with engineering.

Jim: Are there any results to share from that effort?

Gina: They present with me every year. I still talk to five of them. They were the shy girls. Now, they're doing everything. One of them has a fellowship with the New York Hall of Science, following in my footsteps. She was just on TV and a radio show because she created a cool piece of software that has hardware that goes with it. I can't wait to see what they become in college. The program lasted for a year, and it became popular. I begged the principal to let me create an 8th-grade course. I taught the course. We got to go into things like agricultural engineering. We got to do things like create energy. Then, my principal approached me this time and said we need a 6th-grade course and that I needed to pick a teacher, collaborate, and create the course. I chose a teacher I had worked with on STEM lesson units for her class. She wanted to create STEM units for her Regent's curriculum in her earth science class, and she wanted to collaborate with me. So, she became the third member of our team. Now, we present at conferences together. She ran a workshop program for the NYC department of education with me over the summer.

Jim: You said earlier that you used the minichallenges as rewards. Could you say more about that?

Gina: We used the mini challenges because we could slip them into the behavior management system. I wasn't allowed to integrate STEM into the required curriculum. But for me, STEM is the most important thing. I think literacy and math practices can be taught through STEM. I believe that getting students to think creatively is necessary. I don't think we need to teach students content as much as we need to teach them to accept, learn, and share content. We need to show them how to access the things they need to know. In STEM, I teach how to research, how to communicate the results, and how to test different materials or different test iterations of a design to determine success and what revisions may be needed. I teach students how to think outside of the box, and how to come back inside the box when necessary. I teach them to recognize when they need to consider how their big ideas relate to the constraints or the criteria, and that's important. I'm tired of checking off other people's boxes. I've been a teacher for ten years, working with special needs students. I wasn't just a teacher who clocked out at three o'clock. My weekends and nights revolved around bettering myself to teach them.

Jim: Could you tell me more about your relationship to standards?

Gina: I like the Common Core State Standards in some respects. When I finally understood them, I was able to become a leader. It didn't matter that I was a special education teacher. I could go into any classroom and know how to teach the curriculum. I like the Engineering standards, but I don't like content-based standards because I believe that students should be in charge of the content that they learn. It's my opinion, but I think standards should connect to processes, practices, and actions. That's where the CCSS is going. It's getting there. A lot of the science standards are very content-based, and I don't like that.

Jim: That's an important distinction. A lot of teachers complain about the CCSS. What do you think their problem is?

Gina: I know their problem. Their problem is not the standards. It's the implementation of the standards in textbooks, and people's misconceptions of what a standard is. The standards say the child will have a conceptual knowledge of the skill. The publishing companies create worksheets that create confusion about the standards. People think the worksheets come from the CCSS. The standards aren't about making worksheets.

Jim: The Common Core State Standards are performance standards, not content standards?

Gina: Right. They do give examples of content.

Jim: The intent is to have teachers create content?

Gina: The problem starts in elementary school. Teachers who probably weren't good at math are looking at math standards, and they just try to do what the standards tell them to do, misinterpreting it and making it confusing to the kids. The students develop misconceptions because the teachers are just giving them rules to try to make sense of what they don't understand. We have to train teachers and prioritize their understanding of the standards. It took months for us to be able to dissect/discern the pieces that were content and those that were from the process—the pieces that show you how well a student should do something and how a student should do something. There were many implementation problems; for example, graphing moved from 7th grade down to elementary school. Now, we have middle school students who don't know graphs, and teachers blame the CCSS. What happened is that the students were never taught this content because the curriculum moved down to 4th grade when they were in 5th grade. We need to let the new standards go through the system for about ten years, and then, we will see a change. We also need to teach the teachers. They need to learn how we learned.

Jim: What's your relationship to play?

Gina: We do this thing before most learning units called "mucking about" or "messing about." We learned it in the New York Hall of Science. We do a mini challenge that lets students have an opportunity to interact with materials. For example, in the neutral buoyancy challenge, we gave students materials and said create something neutrally buoyant before we asked them to research animals. They were able to touch all the materials they were going to have access to before the real challenge, so they knew the properties of that challenge. We call that play. We also do a play-related activity where the kids are asked to create a tool, but they are making real inventions, and they make it creative and playful. But, I do think in a school setting structured activities are required. The New York Hall of Science has a different opinion on play than schools. Their projects are open-ended. We modified a lot of their stuff to include checkpoints or guidance in some areas so the play could be directed towards a learning purpose. The New York Hall of Science and public school classroom teachers have slightly different goals.

The New York Hall of Science (NYSCI) is a science museum with many science education programs and teacher professional development programs. They host an annual Maker Fair and, according to their literature, are "committed to help bridge the gaps between informal science learning and science learning in the school environment" (http://nysci.org/projects-main/). NYSCI describes itself as a free-choice learning environment that includes in Making, science play, and science career ladders.

Darensbourg and Tesoriero are on the front lines of engineering and STEM education in public schools. They take it upon themselves to learn and create performances of being STEM educators. The STEM focus has transformed who they are in their schools. They have needed to provide leadership on questions of training and curriculum development. They have taken up responsibility for connecting schools to external partners.

From their stories about students and their school communities, they are in fact creating the opportunities for students of color and students living in poverty to be exposed to engineering concepts and STEM activities. They take note of how students respond to the opportunities and have pride and positive feelings about the things they make. Tesoriero discovered that the "low IQ" label that a student may have has nothing to do with his or her ability to create, see differently, and learn developmentally. Student interests and teacher interests can be brought together to create new possibilities. The work that Darensbourg and Tesoriero do in public schools is challenging and can lead to isolation. Their connections to people outside of their schools are important. Those relationships support ongoing efforts and are a source of inspiration. They have discovered that opening the school to partners who are willing to be hands-on is important. They show that innovation and creativity are possible using experiential learning strategies, including hands-on, project-based learning and STEM-/STEAM-integrated curricula. Interdisciplinary perspectives and design thinking are important aspects of the STEAM movement and will enable educators to provide opportunities for children to begin a path toward interdisciplinary learning. Outside partners, such as museums, libraries, universities, and afterschool programs, must play a significant role in this movement.

Some Panel Comments on the Maker Movement

I've created an imaginary panel discussion at an imaginary conference using transcript excerpts from my interviews with Pongratz, Darensbourg and Tesoriero. In the panel discussion they are brought together to share their thoughts on the Maker culture. Martinez and the panelists are sitting in a lecture hall on a slightly raised stage. There is a projection screen displaying pictures of children engaged in engineering activities. The room is filled with undergraduate students and invited faculty members. The topic of the panel is STEAM education, and the discussion is winding down with some closing comments on the Maker movement.

Jim: What do you think of the Maker movement?

Gina: Great question. I think it's fabulous. I believe that it has a place in education, but it needs tweaking. The Maker movement allows students to have fun and create whatever they want. It could be knitting or food or anything. That's wonderful. Teachers are getting involved in understanding what making is. I do see, though, that this open, make-anything-with-anything approach presents a problem when you're trying to do it in the classroom. You need to hold kids to some standards. The NGSS (Next Generation Science Standards) are very open ended as far as teacher choice goes. The NGSS has stand-alone engineering standards for K-12. The DOE (New York City Dept. of Education) has used some aspects of the NGSS in our scope and sequence curriculum guidance. I love the Maker movement, and the design process that's a part of the Maker movement is tangible for classrooms.

Ellen: Whenever you consider the Maker movement or design thinking, STEM is a natural part of that. You have to have that creative aspect to break through. You can't innovate if you don't have a creative mind. You have to have play and creativity, and it has to continue all the way through college because that part of our brain is so important to the STEM field.

Jim: My concern with the Maker movement is that it is only for the privileged classes. What is your take on this?

Christian: We did some community engagement projects in that past that were pretty empowering. We went into my shop and picked up various types of discarded wood pieces, and we took a whole bunch of glue bottles..We were working on a community project and would do installation architecture in the streets downtown. We put fabricated artistic projects on the street, together with music and food. There was a combination of artistic textile installations with many other things, and it changed the street, which was dull, into a very colorful, flowery, environment, where people were thinking differently about where they live. We wanted to involve families and kids, so we went there on a Saturday with our boxes, put up tables, threw all those ingredients of chopped up waste material on the table, put the glue tubes around. We said to the kids, "Now you make something. Imagine something on the street. It can be a building, or it can be anything. Make the street your world. It was fascinating. The kids just spent all day there, working in non-linear ways with non-linear pieces. They used different geometries, but the glue held the pieces together. They were so happy when they finished, and they would put the things they made on the map of the city block and saw the larger picture. I tell you, great things came out of there, and architecture students couldn't have done them better. That's kind of a low-tech fabrication with kids of any age, really; they could have been there all day. It was ongoing. They wouldn't leave.

Jim: It's interesting that you offer that example. I have a similar one. At the All Stars Project's UX program in New York City on a Saturday morning, I led a STEM learning workshop using household paper, plastic and cardboard trash, LED lights (light emitting diodes), rolls of copper tape, and 3-volt batteries. I put everything out on the table, and children, adults, and senior citizens were asked to create objects using all the materials. I demonstrated making a circuit with the LED lights, copper tape, and batteries and asked them to light up their creations. They became fascinated with the technical aspects of building circuits, figuring out how to design switches, and integrating series of lights into their creations. They also loved it. I'm being told we exceeded our allotted time. Thank you, everyone.

Scene ends.

Reference

U.S. Department of Education Programs/Magnet Schools Assistance. (2016). Retrieved from http://www2.ed.gov/programs/magnet/index.html.

CHAPTER 6

Art

Is the goal of STEAM learning to integrate art into science or science into art? The question comes out of awareness that as STEAM becomes part of the curriculum, educators will have to make practical choices about how the integration will happen. Will science teachers integrate a little art into science projects? Will arts teachers incorporate a little science? Will arts and science teachers collaborate to create something new? Who gets to make those decisions and how will those decisions be made? These are all useful content-related and administrative questions that cannot be answered in the abstract. In short, educators will have to figure these things out. The following dialogues offer some new ideas for people who are creating and performing art (or neither) in all types of learning environments.

Performance and Preparation

The following contains excerpts from transcripts of two separate interviews with actors are combined to create an imaginary scene.

Scene:

Martinez, the interviewer, is a middle-aged man and wearing glasses. He is dressed in a dark sports jacket with leather patches on the elbows. He looks like a college professor sitting on a stage in a darkened theater. The spotlights are on, and two empty chairs are angled toward the interviewer but face the audience. Martinez takes a sip of water and then speaks into the microphone, "Can we please have a round of applause for our guests

J.E. Martinez, *The Search for Method in STEAM Education*,
Palgrave Studies In Play, Performance, Learning, and Development,
DOI 10.1007/978-3-319-55822-6_6

Kim Snyder and Marian Rich!" The audience applauds, as Kim, a tall brunette with shoulder length hair walks out onto the stage. She is wearing a smart looking black business suit, a white blouse, and a string of pearls with matching earrings. Kim's is performing as a business executive and she conveys that in her perfect posture and serious expression. Marian follows. She has curly reddish-brown hair and is wearing a broad-rimmed red hat, a sleeveless red gown, and a red feather boa around her neck. As she smiles and saunters across the stage her performance is reminiscent of a cabaret entertainer who sings, dances and tells funny stories. They both sit down. Martinez begins his introductions, and the conversation begins.

Jim: Kim is a professional actress with movie, TV, and theater experience. She is also a playwright. She has had many different kinds of jobs when she's been an "out of work actor". She is currently a corporate event planner. Marian is a comic, improviser, career coach, professional trainer, and owner of her own consulting business, Career Play Inc. specializing in career development. Thank you both for joining us today in this conversation about STEAM education.

Jim: Kim, what was your experience of learning the theater arts?

Kim: I went to a performing arts high school and started working as an actor shortly after. The main thing I learned is that learning is all about preparation. You have to do your homework on your character. You have to read up on the background and the setting, and you have to be credible as a character, which means you have to say and do the things your character would say and do.

Jim: Does that mean you have to have content knowledge?

Kim: Absolutely. You not only have to remember your lines, but you also have to understand a character's motives in a time and place. You have to *become* your character. For example, I played a Native American in a movie set during the Civil War era. I had to read up on the Civil War, what it was about and how that impacted Native Americans. There is a great deal of critical and analytical thinking that is involved in creating a performance that is believable.

Jim: How would you characterize the difference between learning in K-8 school plays and high school and college theater training?

Kim: In elementary school plays, you learn to take and follow directions, *stand there, say this*, and *do that*. You learn to remember your lines and rehearse, prepare to perform. These are the foundations. Later on, it is about technique and critical thinking about character and motives.

Jim: Thanks, Kim. Marian, could you share something about your background in theater?

Marian: My father is an actor. When I was young, my parents played a lot of Broadway musicals on the record player, and I would perform. My father used to give me acting lessons in the living room, and it drove me crazy. I did do theater in summer camp, but when I got to high school, I rebelled against my parents. I didn't do any theater until I got to Sarah Lawrence College. They had a fantastic program, and Wilford Leach ran it. He was a very well-known avant-garde director working at The Public Theatre, and he ran an experimental theater. The students wrote original plays, directed—did everything. You were also required to put in a certain amount of technical hours, so you were involved in hanging lights, building sets. For me, as a performer, it was interesting. In class, I was directed by fellow students who didn't always know what they were doing. It meant I had to figure it out. I had to get on stage and perform my ass off because I wasn't getting that much direction. It was a wonderful environment. I got hooked on theater.

Jim: Marian, Kim emphasized the importance of preparation do you agree?

Marian: Oh yes, it's terrible if you show up unprepared. You can't fake it at all. There's nowhere to hide. One of the things I love about theater is that it's an ensemble. It's fundamental to what theater is. If you're a painter or a writer, it's a more solitary activity. If you're a playwright, to some extent, you probably want someone to read your stuff out loud. If you're an actor or a director, you're always with other people. You're never doing it by yourself unless you're memorizing your lines, which is the hardest part of it, interestingly, because it's a solitary activity.

Jim: Do you think that elementary school plays are of value?

Kim: Absolutely. You learn to speak publicly, you learn to listen, you learn to prepare. Sometimes kids get an opportunity to express themselves in a school play when they are normally considered to be shy.

Jim: Do you think kids could learn some science or math content in a school play?

Kim: Yes. I had a friend who did some work in math learning, and the students performed the mathematical symbols. They had to perform what the symbols did. Also, I think that learning about the lives of scientists and the performances of scientists would be of benefit to students. They would have to learn the language and settings of those characters.

Jim: Marian, how would you approach the integration of science and art or math and art?

Marian: I would want to bring art to science, I guess, or art to math.

Jim: What does that mean to you?

Marian: Well, when I said bring art to math, the thing I thought of was, we'd be working with first grade or kindergarten and painting numbers. Just, getting used to writing, creating those symbols. That's the first thing I thought of, I could see performing certain processes like some kids are going to play hydrogen atoms, and some are going to play oxygen atoms, and two hydrogen atoms and one oxygen atom are going to make water. Then I'd have them perform a skit. The kids are grappling with the concepts, but through performing or drawing or painting or singing. Or even dancing; they could perform a water dance. Or, if we were teaching the planets revolving around the sun, we could create an interpretive dance. That's what I can imagine myself doing.

Jim: How is creativity taught?

Kim: It's all part of the process. You create your character, and you get feedback in the form of critique from your teacher. Your teacher will tell you something like "I believed you all the way up until this point…" Then you talk about what you could have done differently. There's no right or wrong; you have to learn to be in the moment.

Jim: How is teaching in the theater arts different than instruction in other disciplines?

Kim: Well, the teacher doesn't do much talking. There is a lot of practice going on in class. You are either doing a scene or doing an exercise, so you always have to be prepared to be in the moment.

Jim: I know that the important thing in a scene is to listen to the offer and to keep going even when you make a mistake.

Kim: Yes, that's about being in the moment; you forget your line, and you take a beat, and you own it, then it comes to you, and you move on.

Marian: I was thinking of something magical happening. When I teach improv (improvisation), sometimes something happens that is just so magical. For example, I was at this international society for humor studies conference, and we did our Laughing Matters workshop. It was in Ireland, and we had a colleague do the workshop with us. She's a stand-up poet. She did this great warm-up that I loved, where you had to walk around this space and greet people with a curse word, like, "How the fuck are you?" There were a couple of British women who were having a hard time with this. The most extreme thing they would say was, "This is bloody stupid" or "What a bloody rainy day." At the end of the workshop, we had people do skits. One of the women in the debriefing from that warm-up said, "I've never said a curse word. My mother would beat me if I did. We would just never do that. We had impeccable manners."

Jim: What was the magical part?

Marian: The group she was in did a skit about queuing, like at the airport, with your passport and all of that. She was just pretending to read a magazine on the queue. At some point, she put the magazine down, and she went, "Oh, fuck!" It was a magical moment because she did something outside of who she is. It was beautiful. It's not like anybody told her the point of this workshop is for you to curse. It was just a warm-up exercise. How did this woman decide at that moment to be other than who she is? In science, there are accidents and people make discoveries. In improv, creativity often comes in mistakes. You learn, as an improviser to roll with that and embrace that, because that will take a scene from here to there like nothing else will. I think that science, in particular, maybe math too, is taught in such a way that it's about getting it right. If it's only about is getting it right, then you are missing opportunities for creativity, for discovery, and that's got to be problematic in Science because it's problematic in life.

Jim: That's a good reminder that science and math are tools for discovering something.

Marian: Right. I think, for me, it always felt like school was always shutting things down. The science and math part of school seems to limit possibilities rather than opening them up.

Jim: Marian and Kim, thank you so much for joining us. We only have a few minutes left. I want to open up the conversation to the audience. Are there any questions?

End of scene.

The distinctions that we typically make between performance on stage and performance in life cease to make sense if we look at how the tools of performance can be used to create new ways of being in life. While Kim and Marian do not have backgrounds in the sciences, each seemed open to the idea of creating new performances with science and math. They both performed the improvisational "Yes, and" to the questions. They did not disqualify themselves from the performance of an answer. They were each able to draw from their life experiences to respond "in the moment." I can contrast this with the experience of working with higher education faculty trained in the sciences. The examples of teaching abstract math and science concepts with art make the concepts more concrete and approachable. Many of my colleagues unintentionally marginalize the idea of using performance and play in teaching. There is an insistence that learning activities must be related to meeting accreditation requirements or learning standards. I would characterize this as being unable to create "being in the moment" in formal learning settings. It is possible to use predetermined tools without being predetermined by them (see Part II, Learning and Development). Some people see acting/ performing as "pretending" but I (and these women) see it as "transforming" "being and becoming"; this what I try to create in classrooms. I see learners *like actors who become their characters, students in performatory learning environments become scientists; when they are working on a chemistry lab or figuring out how to build a suspension bridge, they are doing the work of and becoming chemists and engineers.* The next dialogue will continue to develop ideas on the relationships between science, art, and technology in learning.

A Playful Interdisciplinary Artist

Yuko Oda is a visual artist and professor of fine arts in New York who teaches art in the College of Arts and Sciences at NYIT. Oda's animations, installations, sculptures, and drawings have been exhibited at many art galleries in New York and China. She has received awards for her work in animation. Oda earned her M.F.A. from Rhode Island School of Design.

Jim: Could you say a few words about what you do?

Yuko: I am a practicing and working artist in New York City, and I have been teaching for ten years now in the digital art and design department. My specialty here has been computer graphics and 3D modeling and animation. I see myself as a visual artist. I have animation, 3D, and sculpture as part of my practice; I'm not an industry animator. Regarding my visual art practice, I see it as very experimental and exploratory, where I express myself in a variety of media. I got my undergraduate degree in painting and my graduate degree in sculpture studies. I started teaching in a visual media field 14 years ago. My work right now is in three areas, which are 2D paintings, drawings, and collages, which is already multimedia. I also do digital sculpture, which is 3D printing mixed with installation art, and the third area is animation, using a variety of different animation media. What I like to do is dabble in all types of expression. But the visual message and my style remain consistent even though I use a variety of different media. Does that make sense?

Jim: I don't know what "your style" means.

Yuko: Ok, for example, there are consistent motifs or markings in my work. Let's just take something like a dandelion seed flower. You will see that in my 2D drawings, you will see it in my collages, and you will see it in my animations.

Jim: So, anytime I see something that you've created, I can look for that and probably find it someplace.

Yuko: Yes, you can see something I've created and think, "Oh, this is Yuko's work." There is something recognizable about it.

Jim: And you do that, for what purpose?

Yuko: I think because it is part of my expression. It remains consistent without me even trying to make it consistent. I love the fact that I am in multimedia, but it can be very confusing, too, because I have a whole toolbox of a variety of ways to express myself. There is a push and pull to wanting to try and explore different media, but also to focus on using the same medium over and over again to hone my skills with that particular tool. Part of me wants to paint on the same size canvas, and there are artists who do that; they paint for 30 years on the same 12" x12" canvas, and their work always remains in the same format. I look at that, and I think, "Wow, this person is a real expert at painting in 12" x12" canvas format." But I find that quite boring. I want to explore a new way of using different materials each time. I find that the play that's in my work, with materials, is part of my exploration as an artist.

Jim: You're playing as you're performing as an artist?

Yuko: Exactly. If I don't have that playful spirit, then I don't feel like I'm making valuable artwork. So, I feel like it connects to that first question we asked ourselves (before we started recording). Are we children or are we adults? I will always have that child-like, play-like spirit in my work. That probably means I won't stay with one medium, but I will explore a variety of media.

Jim: And that tension you feel, when you were talking about the push and pull?

Yuko: I feel, in some ways, the evolution of my work and the exploration of media can happen quite quickly. There is also a part of me that wants to stay put in one exploration and push that further. I don't think there's a wrong and right process. Maybe this happens to scientists and educators as well. You want to research and stay on one topic for one extended period, but then you also become interested in another topic. How do you stay fresh and motivated with a variety of research topics if you just stay with one?

Jim: I think this is amazing. What do you do about the push-pull?

Yuko: Sometimes, when I'm working on a piece of art, and I have a piece that I need to finish, I will have many ideas that come up in my mind. Part of me wants to abandon what I'm working on to pursue the next idea. What I do is jot it down in my sketchbook—a visual sketch or just jot down a sentence or two of what I know it is—that way I can still keep that idea, but finish the piece I am working on. That's one strategy I have for staying focused and following through with a project.

Jim: I'm trying to figure out if what you're expressing is just a normal part of the process of becoming an artist. Is something else influencing your ideas on the work?

Yuko: There are two answers to that. One, I think it is intrinsic to the artist, that whole process of play. Two, I think I have big influences that inspired me to work in that way. I used to be a kindergarten art teacher right out of undergraduate school. Kindergarten art teaching is offering the child a variety of materials, and saying, "Okay, play." Young children are experts at playing anything. They can make an art project out of a recyclable object. Usually set up takes 20 minutes, playtime is 15 minutes, and cleanup is 45 minutes. It takes so long to clean up afterward, but it was so opening and fun to be in that role of facilitating that play. After I had taught kindergarten art for a couple of years, my artwork changed drastically, and I started creating installation art out of recyclable objects. For 10 years in my art career, I created installation art out of objects that we use as disposables.

Jim: "Installation art" means?

Yuko: Installation art means using art from the space around you. For example, you might suspend things from the ceiling or place things on the floor. I should show you some images of this. I did a whole installation out of bubble wrap. It looked like a children's room or playhouse, but everything was made of bubbles and air. That work got me into the best art schools in the nation. When my paintings from before were getting rejected from some of these schools, the bubble wrap art got me into the art Institute of Chicago, Rhode Island School of Design, Hunter College, some of the best MFA (Masters of Fine Arts) programs.

Jim: Working with children helped you get into Art school?

Yuko: Well, yes, teaching them helped free my mind to spontaneous play, and I think that that allowed me to find my voice as an artist. The second influence I can think of is Donald Greenberg, one of the speakers from the Siggraph Art Conference. That's the computer graphics and interactive techniques conference in the U.S. There's also a Siggraph Asia, which I'm participating in December in China. When I went to Siggraph in 2009, in Los Angeles, I saw Greenberg, and he talked about his process as a researcher in computer graphics. What he allowed himself and his team to do is to find inspiration and topics that seemingly did not relate to each other, but his research method would start from something that fascinated him, like a bird's wings, and how the bone structure of the bird's wings moved. He would study that, and somehow that would lead to how to create a prosthetic leg for a human. The way he researched had so much freedom to it. He went from one topic to another that seemed not necessarily to relate, but in the end, they did end up informing each other. I guess it's allowing your instinct and intuition to move freely between disciplines. It's computer graphics and art and programming, and so it allowed something unique to happen because it was interdisciplinary. After his talk, I felt like I shouldn't be so hard on myself. I should embrace that and trust my instinct instead of trying to pigeonhole myself into one category.

Jim: I think the fact that you are saying these things about art and play means that we are onto something regarding the nature of being interdisciplinary. Isn't this contrary to how we organize schooling?

Yuko: Well, as an artist, I feel like I have to be a scientist. For example, when I'm figuring out how to put together a collage, I have to understand what type of glue works for a particular kind of surface. It's trial and error. I have to try seven different kinds of glue to figure out how this one plastic will adhere to a particular Japanese paper. In that sense, there is a chemical reaction that happens that I have to observe through trial and error. I feel like when people in an educational system think of science, they believe that it's got rigor compared to art. Some people think art is just fluff, and I don't believe that's true. There is a value in both.

Jim: I agree with you, we have achieved a lot in science, and you can't argue against the scientific method. Isn't it a different kind of creativity that we're talking about here?

Yuko: Right, you need both. One thing that Greenberg was talking about at the Siggraph conference is that it's valuable for a scientist to learn an art and an artist to learn science. Or for an artist to learn computer science to be more innovative. It used to be that artists just did one type of thing, and scientists did just their thing. But now, we have to train ourselves to do both.

Jim: What do your collaborations look like now? Or do you collaborate now?

Yuko: One area of research I've been doing is with a computer scientist, Ted, a software developer at AT&T. On his side projects, he likes to collaborate with artists. He and I were creating some games where he would program and I would create the visuals and the concept of the work. That was more of a computer game multimedia type of thing. We got to show it at the DUMBO arts festival a couple of years ago. I couldn't have done it without his help. Another collaboration was in 2015 with a French singer/songwriter. She created the music and song, and I created an animation to go with it. She performed the song and had my animation playing in the background at a variety of venues in New York and Paris. We want to do more together. We are still brainstorming ideas.

Jim: What happens when you don't know what your collaborator is saying about the technical aspects of the work?

Yuko: In the first project I was talking about, Ted, the computer programmer, is always sitting next to me typing away at the computer language. I've taken some programming before, but I don't always understand all the complexities of the programming. I don't know if that answers your question, but I felt like there was that gap of his language and my language of a visual artist.

Jim: Is there a negotiation process?

Yuko: I would tell him that I wanted the organism on the game to move in a particular way, to pivot or rotate in a certain way. He would type that formula up in the language, and then we'd execute it and check it out. I would visually test what he did, and I'd give him feedback, and he'd fix it again. He would also explain to me what he was doing, so I could read it and understand it, but I couldn't create the language as he could. Does that make sense?

Jim: Absolutely. What I hear in the collaborations that all sorts of people have is that there are these invitations to make new meaning together, because you are doing something that you couldn't make without the other person.

Yuko: Right, and Ted couldn't make the visual and the concept on his own, so he's gaining by having that experience because he gets to put together what is very visually compelling and engaging. My second example, where I made an animation with the musician, that was really interesting because of the musician. She made a very spiritual and uplifting song. I didn't listen to the lyrics. I just lost myself in the feeling I would get from the song. The animation was inspired by how the song made me feel. Then, when I showed her the visuals to go along with the song, she cried. She said, "this is the most beautiful animation." I showed her maybe one minute of it because it takes very long to animate, so a minute takes forever, and it's a 3-minute song. She told me she wanted me to continue, so I did. There was no clear storyboard or clear formula of what she wanted me to create in animation.

Jim: You were creating together without a predetermined idea?

Yuko: I didn't know what was going to come out. In this piece, I go from a stop-motion animation of dropping acrylic ink drops onto pieces of paper and taking pictures to make the visuals to a 2D animation using after effect. Then at the end, it breaks through to a 3D animation using 3D modeling. So it goes through those media I was talking about before, all in one piece. Once again, I'm breaking through those boundaries of categorization. I think that's part of my artist statement. It's part of the process that I like to play with. It's to explore and break through those media, and I couldn't get that piece without her music. I wouldn't have been able to create it because that inspiration I felt from listening to her music was what gave birth to that piece.

Jim: How do you support the students to be creative on school assignments?

Yuko: While being able to assess them?

Jim: Yes, how do you negotiate all of that?

Yuko: It's a broad question. I think students do best when they are first taught how to use the tools and to the best of their ability learn to practice with the tools, so they have a tool set that they can express themselves with. What I teach in my class is quite technical; it's 3D modeling and animation. I try to have them familiarize themselves with the technical aspects of the tools. But then, with the assignments and projects that come up, I like to give them creative freedom.

Jim: How do you get them to be creative?

Yuko: I tell them that the most important thing is their artistic voice, what they want to create, what they want to say as an artist. There are specific exercises to build skills where they have to follow the step-by-step instructions and create something that looks exactly like some example. That's not creative, even though you're creating something, because you're following the steps. But then, after they learn how to create by following the steps, they have to come up with their own creative ideas. For example, one project I have for the introductory course is that students have to create a vehicle of transportation that doesn't exist right now. We research a variety of different vehicles. For example, it can be anything from spaceships in Star Wars™ to the house that flies away in *Up*, the Pixar™ movie. We talk about why that house is a vehicle and how it's convincing even though it's a fantasy vehicle. In art, you don't necessarily have to follow the rules of engineering or architecture. As long as it's convincing to our imagination, we can make something meaningful.

Jim: Which is how we created most of the engineered and built things in the world.

Yuko: Right. I think it starts out with a visual idea, and then you have to figure out how it's going to stand, etc. I want students to create something that is unforgettable and truly innovative and describes a limit regarding what type of vehicle it is. It can be in outer space, it can be something inside your body, it can be an abstract vehicle, it can be driven by aliens or humans in the future or by ants underground. The students get excited because they can create something themselves. They are not what told to do. We spend a whole week just on their idea development. I emphasize the grade of the creative project over the grade of the project where they just follow directions. I remind them that their ideas and creativity are what's important because everyone can recreate what already exists, but what is unique and original to the artist, no one's created before.

Jim: Do undergraduates become uncomfortable when you ask them to do this?

Yuko: Yes. All along, they've wanted the freedom to make stuff the way they want to make it. Then they get to the 3D class, and I say, "You have all the freedom in the world, come up with the idea," and a lot of them freak out because I'm not telling them what I want. Right now, with my senior project class, they're doing a similar thing, but they are creating characters. I said, "The character could be anything you want, as long as it's yours." We spend a whole semester building the bone structure and bringing life to the character, and then it will do things at the end, like walk or pose. It's like Frankenstein; we build life into the character.

Jim: What's the hardest part for the students?

Yuko: The hard part is coming up with the idea. Students can't wait to get to the characters, but then they get to it, and they're like, what is our idea? I ask them what kind of characters inspire them? We start from there. If they don't have ideas, we look at a lot of references to see what inspires them. I remind them if they have this as a semester project and it's not something they're inspired by, it's going to waste their time. It's through these projects, where they have to create something from their imagination, that they connect to that artistic voice. It's 3D modeling and animation and using technical software, but really, it's about the artist within.

We discovered in conversation the dialectical unity of Oda's artistic process of discovery, the work and the play. She leads her students away from following steps in a process for the purpose of developing the skill to a process of discovery that encourages them to engage in the uncertainty of creating something new. There is no separation in her process between the science and the art, and each creates activity in the other domain. The animation would be impossible without the computer science, and the computer science is without artistic purpose if an artistic voice does not emerge.

Oda's experience as a kindergarten teacher is fascinating because she provides an example of the kind of development that Newman and Holzman describe as being a part of social therapeutic practice. Everyone has the possibility of development, and everyone can contribute (Newman and Holzman 1997; Holzman and Mendez 2003); Yuko needed to know something about art and teaching to create a developmental learning environment for children. The children did not have to know anything about art, teaching, or learning to create an opportunity for Oda to learn and develop artistically. Oda's interaction with young children provided a chance to experience how play and art and not being predetermined by the process come together in a creative activity. Somehow her formal training in art did not provide her with the opportunity to develop her artistic vision. There is no explanation as to why Oda experienced her artistic transformation in teaching kindergarten. It's important to understand that no explanation is needed. Oda has discovered an important insight on learning and development.

Vygotsky, Creativity, and Art

Vygotsky was interested in art and psychology. According to N. Cathrene Connery (2010), Vygotsky's dissertation attempted to address the psychological questions related to art. He was interested in the potential of art to be curative of psychological problems, and "…he proposed that the foundations of art are rooted in the human need to manage and release intense physical, mental, or emotional strain" (Connery 2010, p. 21). Connery also notes that "Vygotsky observed that artists and audiences alike can achieve the transcendent resolution of emotion by engaging in the reciprocal processes of creative production and aesthetic response" (p. 23). Connery, citing Vygotsky's dissertation, notes that art "opens the way for the emergence of powerful hidden forces within us" (p. 26). Vygotsky's early work resonates with our discussion of art and the powerful force of "artistic vision," creativity, and learning in ways that are not so obvious when reading his later works.

Oreck and Nicoll (2010) note that there are "complex relationships involved in the development of dances and dance artists" (p. 108), citing Vygotsky's dissertation, which asserts that "[t]he act of artistic creation cannot be taught" (Vygotsky 1971, p. 256). Oreck and Nicoll focus on understanding the role of the teacher in the zone of proximal development. They cite the work of Henry Schaefer-Simmern (1948) who noted that the "teacher's job is to facilitate the individual's 'awakening.'" This idea about teaching art bears a resemblance to what Yuko offers when she works to develop the "artistic vision" of her students. Oreck and Nicoll explain the struggle in the artistic classroom in the following observation: "The conventions of teaching through demonstration, modeling, and scaffolded instruction—often by breaking the whole into parts—offer structures to guide learning but may also reinforce a student's tendency to follow rather than initiate or innovate" (p. 109). They also provide some support for the role of play in artistic process with their observation in the realm of dance instruction that "[d]irected physical play helps artists resist the tendency and pressure to intellectualize the process" (p. 115). The pressure to intellectualize connects with Yuko's description of having a "playful spirit" as a counterpoint to the intellectual work that she does.

Lois Holzman (2010) citing Vygotsky (1979, pp. 102–130) notes that Vygotsky made a distinction between the zone of proximal development (ZPD) of "play-development" and the ZPD of "learning-instruction development." Vygotsky's distinction focused on play being the "highest level of preschool development." Holzman suggests that difference does not need to be as sharp as Vygotsky describes it and observes that Vygotsky "may have overlooked some continuity between the two ZPDs, in part because he was so concerned with learning in formalized school contexts."

In the closing remarks in the edited volume *Vygotsky and Creativity*, editors Ana Marjanovic-Shane, M. Cathrene Connery, and Vera John-Steiner offer the following critique of education. "In our view, the American public educational system suffers from a serious lack of vision and action when it comes to the development of creativity. The very structure and practices of our K-12 system restrict, retard, or prevent imagination, play, and creative ingenuity across the disciplines" (2010). I would add that maybe the vision that the educational system is lacking is a performatory vision.

Questions and Answers

Is the goal of STEAM learning to integrate art into science or science into art?

Jim: The question places the product (goals of STEAM learning) ahead of a process of learning. A learning goal is not the starting point to a creative process, it's something that might be achieved. The dichotomous formulation of art or science ignores the continuity that exists between the disciplines. That continuity can be rediscovered in play and performance

Will science teachers integrate a little art into science projects? Will arts teachers incorporate a little science? Will arts and science teachers collaborate to create something new?

Jim: The answer to the first two questions is yes. The answer to the third question, to paraphrase Newman and Holzman (1997), remains to be performed

Who gets to make decisions about integrating arts and science, and how will those decisions be made?

Jim: My suggestion is to let students guide and make all those decisions. I suggest performances starting with learning to say "yes, and" to students and building from there. For more information about performance and improvisation, see Lobman and Lundquist's *Unscripted Learning, Using Improv Activities Across the K-8 Curriculum* (Lobman and Lundquist 2007)

References

Connery, C. M. (2010). The historical significance of Vygotsky's Psychology of Art. In M. C. Connery, V. P. John-Steiner, & A. Marjanovic-Shane (Eds.). *Vygotsky and creativity: A cultural-historical approach to play, meaning making, and the arts* (pp. 17–25). New York City: Peter Lang.

Holzman, L., & Mendez, R. (2003). *Psychological investigations: A clinician's guide to social therapy*. New York City: Brunner-Routledge.

Lobman, C., & Lundquist, M. (2007). *Unscripted learning; Using improv activities across the K-8 curriculum*. New York: Teachers College Press.

Marajanovic-Shane, A., Connery, C. M., & John-Steiner, V. (2010). A cultural-historical approach to creative education. In C. M. Connery, V. P. John-Steiner, & A. Marajanovic-Shane (Eds.). *Vygotsky and creativity: A cultural-historical approach to play, meaning making, and the arts*. NewYork: Peter Lang.

Newman, F., & Holzman, L. (1997). *The end of knowing: A new developmental way of learning*. New York, NY: Routledge.

CHAPTER 7

Math

School boards and school districts determine the mathematics that we learn in school. In Chap. 1, the NCTM (National Council of Teachers of Mathematics) was introduced as a major stakeholder in math education. The NCTM is one of the primary influencers of math standards and works with textbook publishers. The math standards are used by publishers to outline what students should be able to do at the end of a course of study at each grade level. The textbook publishers also work with the professional societies and local governments to create the standardized assessments that are used to determine whether or not students have the knowledge that they are expected to have at the end of a learning unit. Federal- and state-level legislators also influence what happens in the mathematics classroom. One mechanism for the federal and state governments to track student performance in math at the national level is through assessments administered by the National Center for Educational Statistics (NCES).

The NCES publishes the National Assessment of Educational Progress (NAEP). This annual national report provides assessment data on the performance of tens of thousands of American students in various subjects. The NAEP Trends in Academic Progress report provides trends on reading and math assessments dating back to the 1970s (US Dept. Education 2013). The following analysis of results in mathematics knowledge is offered: "[r]esults from the long-term trend assessment show improvement in the mathematics knowledge and skills demonstrated by 9- and 13-year-olds in comparison with students their age in 1973,

J.E. Martinez, *The Search for Method in STEAM Education*,
Palgrave Studies In Play, Performance, Learning, and Development,
DOI 10.1007/978-3-319-55822-6_7

but no significant change in the overall performance of 17-year-olds" (US Dept. of Education 2013, p. 6). This means that over the course of 43 years in mathematics education, we have implemented changes that have improved achievement levels for 9- and 13-year-olds. However, those gains do not translate into higher performance by the time students are 17 years old. Other positive news offered in the report is that gender gaps and racial gaps have narrowed, but not consistently across all subgroups.

It is hard to believe that 43 years of reform in mathematics education has not resulted in a significant difference in the performance level of 17-year-olds who will soon enter college or the workforce. However, there have been significant increases in spending over 40 years, and there have been changes to curriculum, standards, and teacher preparation. Teaching and learning in mathematics have changed, so perhaps we need to take a different look at how we teach math and evaluate the outcomes of instruction. I don't hold out much hope that the data-driven instruction (using assessment data to make instructional decisions) strategies that have been popular since No Child Left Behind 2001 will lead to transformative change in mathematics education. I do have hope that performatory methods that are embedded in learning standards can become prominent in the practices of creating interdisciplinary learning environments.

Performances of Math Conversations

I recall standing in the back of a middle school mathematics classroom in 2008. I was in the back of the room repairing laptops in a laptop cart. The tables and chairs were arranged in a large U-shape with the open end of the U pointing to the front of the room where the old blackboard and the new electronic whiteboard were. Beth Smith (a pseudonym) the teacher, moved around the room and worked with groups of students. The students were seated on the inside and outside of the U, facing each other. The classroom was noisy, but the students seemed to be talking to each other and working. There was nothing about what was going on in the classroom that was unusual. I had been in Smith's room on many different occasions, and I knew all of the students. We were a small school. I was the Tech Teacher, and fixing equipment in the classrooms of my colleagues was part of my weekly routine. Group-based learning was a norm in our school; students in grades 6–8 all worked collaboratively. A significant number of these middle school students were from affluent White families in lower Manhattan, but there were also many students who were Black, Hispanic and Asian from working-class and low-income families. Students had to go through a competitive application process to get into the school.

On this particular occasion, I heard something that drew my attention. Smith had asked the students to stop working, and she wanted to listen to their conjectures. Their task had been to find patterns in numbers and generate theories and develop proofs. The students started talking about their conjectures using the math vocabulary that Smith had been teaching them. I stopped working on the laptops and just started listening to what the students were saying. I was fascinated. They were describing properties of numbers that I had either forgotten or never learned. Some of the students had misconceptions that I recognized, but Smith only asked questions and did not indicate whether a student said something that was right or wrong. Some of the students debated with each other about their conjectures and proofs. Instead of declaring someone correct, she asked the group whether there was a consensus. This episode was not the only one that had surprised me in Smith's classroom. On another occasion, Smith requested the installation of specialized graphing software on the laptops. The students were using graphing calculators that had data ports that would allow them to save work on laptops so they could print graphs, manipulate the formulas interactively, and change plotted curves and compare them in real time. It had been a very long time since I had been in middle school, and this was all new to me. Even my college-level math courses in algebra and statistics had never provided me with opportunities to discuss proofs and debate about my conjectures. Things had changed in math education since I'd learned math in school in the 1970s and 1980s. math, at least in 2008, had become a social and hands-on technical activity, and I think that this was a promising development.

Smith and I shared some things in common: We are both career changers and had experience in corporate settings, and we both went through the New York City Teaching Fellows program that provided us with alternative routes into teaching in the late 1990s and early 2000s. Smith had worked at a large consulting firm and had a background in theoretical and applied mathematics. Unlike many math teachers, Smith had been a professional mathematician and was a gifted math student in school. When I watched Smith interact with some of the talented math students, I could understand that something different was possible with

Smith as the teacher. Smith was not just pushing gifted students. She supported students who struggled to stay in math conversation with gifted students. To put it in Vygotskian terms, Smith created zones of proximal development (ZPD) that had different possibilities for different students. The same opportunities did not exist with other teachers, but this was not because Smith was a math genius. Smith created mathematics conversations in her class, and she provided all students with the performance directions they needed to create convincing performances of talk about math. The vocabulary, style of spoken language, and math concepts were important in creating performances of talking about math.

Smith led the students to discover concepts, patterns, and facts. She encouraged them to imitate how she led group conversations. Mathematics became a complex social activity, something that everyone could be organized to do to discover how math was used in culture. In Smith's class, it was possible to perform math learning. What Smith created was a developmental environment for math learning by creating opportunities for mathematical discourse that were grounded in social activity, not in the solitary mental effort of figuring out how to get the right answer. She was encouraging what Holzman refers to as "math talk" (Holzman 1997, p. 122).

Smith's approach to math talk in the classroom is described in the Common Core State Standards for Math (CCSSM). The standard known as MP3 reads as follows: "Construct viable arguments and critique the reasoning of others" (http://www.corestandards.org/Math/Practice/MP3/). The performance of math talk I witnessed is described by the standard. I've always considered Smith's class to be a developmental learning environment, and the standard describes what I observed. Following the standards should help a teacher in creating developmental math learning environments. Unfortunately, there are many challenges to creating developmental learning environments in school.

It Is Hard to Be Developmental

Vincent Accardi is a mathematics teacher in New York State who has been teaching math using the Common Core State Standards, and he is also a student of mine in a graduate course. Below are some of his posts in an online course (edited for clarity). He shares some of his thoughts on why math is hard to learn in school despite standards that encourage developmental approaches:

Post 1

Why do so many people find mathematics cumbersome and, more so, particularly challenging? The goal of teaching is not for students to memorize facts and formulas because that has little to no value in the outside "real world." Rather, we wish that our students can conceptually understand the material and eventually be able to manipulate their conceptual understanding of the mathematics into something that is of value to them when they are adults. In the math classroom, students attempt to understand the problem before trying it. Students use drawings, graphs, and physical models to help solve problems. We teach them appropriate strategies for solving different types of problems and much more. As much as we don't want to admit it, we are teaching to the test. With tests come deadlines and due dates, and with this also comes pressure to move through the curriculum at a steady pace, and sometimes this leads to us to the need to just move on regardless of student readiness.

Post 2

I agree that the math classroom can produce competition. Collaboration is something that I tried to stay away from in my education. There are always winners and losers in the classroom, and I used to be a firm believer that competition in the class was the best way to create self-motivation. I was wrong. The playing field is never going to be fair, there are always going to be students that are well ahead, at, or below grade level, and this is simply not a situation where competition should take place. The "losers" can be identified before the game even starts. As a competitive student, if I thought that the teams were too uneven, I would simply not play. Some students must feel the same way when they enter a classroom, and it is so sad. Collaboration and group success are what I celebrate now. When students in my class who are "winners" try to answer the problem before I finish writing it on the board, I offer them extra credit as incentives to provide substantial help to another student.

Post 3

I am always attempting to find ways that will help me convince my students that mathematics is more than one isolated topic taken out of context. I receive a healthy dose of "When in God's name will I need to know how to (fill in the blank)?" And the answer is commonly, *never*; a student will almost certainly never need to perform a dilation followed by a reflection across the line $y = 4x$. The purpose of most of these standards, for me at least, is to develop or enhance the mathematical competencies and conceptual understanding. "Conceptual understanding" is a phrase that's tossed around. I believe that most students don't have all the conceptual understandings they should. In my estimation, a majority of students are graduating with a memory of procedural fluency that will almost certainly be lost within a few months.

Accardi is using the CCSSM, and those standards give shape to what he does in his classroom. He identifies teaching to the test, competition in math learning, ideas about collaboration, the need to cover material at a certain pace, teaching learners at different levels, and the relevance of the content as part of the everyday challenges he takes on. Accardi's self-identified movement from being competitive and not collaborating to encouraging collaboration and his recognition that competition is not appropriate in a classroom that is "not a level playing field" are signs of his development as an educator. His posts indicate that he has developed empathy for his students, acquired through experience as a teacher and building relationships with students to create new possibilities in the classroom.

Brains Pre-Equipped for Doing Math in Social Settings

Despite our difficulties in learning math at school, math concepts are as available to human beings and language is. We are all born with what is referred to as *number sense*, and we do not need language to experience number. Research conducted by Gallistel and Gelman has led them to compare human and non-human cognitive mechanisms in number sense. They have discovered that animals also have "a non-verbal system for representing discrete and continuous quantity that has the formal properties of continuous magnitudes" (Gallistel and Gelman 2005, p. 31). The researchers assert that non-verbal tools for mathematical reasoning develop at about the same time that children become aware of the world. We have number sense, a capacity for numeracy that is independent of language and knowing that symbols like 1, 2, and 3 mean a quantity of something. However, numeracy is situated it does not get put to use as an independent cognitive function in the brain.

In the 1980s, Jean Lave conducted studies that contrasted the abilities of shoppers to do mental mathematics calculations while shopping with their abilities to do the same calculations presented as math word problems written on paper (Lave 1988). Lave demonstrated that shoppers were using the setting, the supermarket, in ways that materially contributed to making the mental calculations. We can frame the shoppers' mental computation process as an activity that was in dialectical relationship to the supermarket. In other words, the setting influenced the relationship of the individual to the mathematical activity in complex ways. According to Lave, the shopping setting provided the possibility for choice making, revisions to calculations, meaning-making, comparing estimates, and simplification, all of which contributed to improving mental calculation accuracy in routine shopping. Lave concluded, "arithmetic is more structured by than structuring of shopping activity" (Lave 1988, p. 158). Simply put, how arithmetic is done is influenced by shopping.

In contrast, a formal, school-based testing environment does not provide opportunities for choice making (a sense of control and autonomy) or meaning-making. Lave found that shoppers who considered themselves poor math students in school performed better in the supermarket than in conditions that felt more like a test. They were unaware of their efficacy in the supermarket setting (Lave 1988, pp. 168–169).

Barbara Rogoff, like Lave another pioneering Vygotskian, provides further support for the idea that human cognition (including mathematical cognition) is situated in social practice (Rogoff 1990). She offers a framework that highlights the active role of children in making use of guidance, the importance of participation in routine cultural activities that are not instructional, and cultural variations in goal and means of engaging in a shared understanding of activities with those who are guides and companions (Rogoff 1990, p. 8). According to Rogoff, these guides could be peers and adults who explain, discuss, model, observe, and influence how children take roles in cultural activity. Rogoff stressed the idea of *intersubjectivity*, the shared purpose and focus of children's interaction with skilled partners as the underlying process that supports children in appropriating skills and increasing understanding (p. 8). Rogoff makes it clear that cognition and cultural, technological, and intellectual tools used in activity are all tied together. She uses examples from different cultures to point out how "mathematical tools are an essential aspect of numerical thinking, and how individuals' skills are unique to the tools they have used" (Rogoff 1990, p. 52).

Michael Cole, a Vygotskian, in partnership with the Distributed Literacy Consortium studied the effects afterschool settings in math learning (Cole 2006, pp. 93–98). They found that children who participated in afterschool programs (Fifth Dimension) that featured the participation of college students engaged in shared activities of playing games and puzzles on computers demonstrated significant differences in "end of grade test" scores to control groups. The Fifth Dimension children's afterschool programs, based on Vygotskian theory, were found to have a positive impact on children's academic skills. However, the researchers were unable to provide direct evidence of the kinds of interactions that were producing the positive outcomes (Cole 2006, p. 107).

The research cited above suggests that mathematical reasoning is as social and as natural to human beings as our capacities to use language and technology. If that is the case, then why do we believe some of us can do the math and some us cannot? Creating more opportunities for "math talk" may help students and teachers create new math performances in classrooms. I think the positive and socially active environments that teachers like Smith and Accardi can create go a long way toward helping students believe they are capable of learning math. Accardi indicated that one of the social forces in the math classroom is competition. It has the unfortunate effect of reinforcing the beliefs in students that they cannot compete. However, this is not the entire story.

A student who achieves 100% on a test receives positive reinforcement and is expected to maintain that level of achievement. Students who receive a grade less than 100 are challenged to do better. The testing system competitively ranks students against each other. The Vygotskian scholars cited above are clear on the complexities of the social environment, including opportunities for social and emotional development, as being factors in mathematical cognitive development. Therefore, we have to be open to imagining math education in a socially complex environment that is not reducible to the score a student gets on a test.

Students ask questions about the relevance of math instruction throughout their K-12 experiences, with many math teachers admitting that there is not much relevance in much of the content to the lives of students. When mathematics is part of social activity, including shopping, play, and puzzle solving, the question of relevance disappears. Another Vygotskian researcher, Mike Askew, director of BEAM Education and a professor of mathematics education, suggests that creating meaningful contexts for solving problems is important for supporting children to "rise to the challenge" (Askew 2008). Askew is a proponent of math talk in the classroom and suggests providing students with opportunities to rehearse in the classroom prior to being asked to present (Askew 2011).

Askew is also deeply involved in teacher training and uses the metaphor of a teaching tripod—task, tools, and talk—to organize useful contexts for teaching math in the primary grades (Askew 2016). The beliefs of math teachers figure prominently in their effectiveness in the classroom, and effective teachers encourage discussion, use students' descriptions to emphasize connections, and build on students' mental strategies for calculating (Askew et al. 1997). Finally, an important step in creating developmental, interdisciplinary math-learning environments will be to challenge misunderstandings and oversimplified assumptions about students.

Performatory Tools

Part I of the book contains an anecdote describing math learning in a playful, technology-rich environment where students were performing their mathematical roles in the math Video Project. They created meaning using math, technology, and artistic media such as songs and animations. Early childhood math-learning researchers like Nicola Yelland connect positive effects in learning math with technology use in the classroom (Yelland and Kilderry 2010). The Math Video Project is a relational "tool and result" type of tool. What I have heard Holzman say on numerous occasions is that we are interested in creating "new wants." I believe this is important and speaks to creating relevance in mathematics topics that do not connect to everyday life. The math Video Project created a new kind of math challenge in daily life, and the students wanted to show up to Friday's math class to make math relevant to everyday life by producing a video that anyone could watch. The project created "new wants," which should not be confused with what teachers would refer to as motivation. New wants are generated in activity, whereas motivation is a tool that is used to produce a result. In the next section, we meet Jeff Lisciandrello a math teacher who created new wants for himself and his students, with assessment and training technology in the math classroom.

An Interview with Jeff

The following dialogue is from a meeting with Lisciandrello, a former student of mine. He was a fifth-grade math teacher in New York City. He describes his experience in using technology to transform math learning and assessment in his classroom. The dialogue has been edited for clarity.

Jim: I want to talk about one of the projects you did with the Khan Academy™ web-based assessment tools with a couple of students in your class. Can you set the stage?

Jeff: When I began teaching math, my students were given a diagnostic, and almost every student failed. I assumed they were ready with core math skills that they would have developed in fourth grade—things like adding three-digit numbers, multiplication tables, recognizing place value in decimals. A lot of students were not comfortable with that level of math and did not demonstrate mastery. When we moved into the first chapter of the textbook, I realized there were so many gaps, we could not start with the textbook. I looked for alternatives, and the director at my school recommended Khan Academy™. It's a system that generates a personalized diagnostic for each student. The interactive software prompts the students with questions and determines the grade level of the student. The system would automatically give students question items that were suited to their level. The software generated data that I could use. It created visualizations of that data that were easy to understand. Even for somebody who is not into collecting, organizing, and analyzing data, it does a lot of the work for the teacher. All a teacher has to do is read a graph and he can get a lot of information about where each student is and where the class is as a whole. It gave me a level of assessment and a level of understanding of students' individual skill sets that would have been impossible for me to achieve with just paper tests.

Jim: You created some learning interventions for two students. Would you describe what they were?

Jeff: Those two students were several grade levels behind. In addition to that, they were acquiring skills more slowly than a lot of other students in the class. What we did with the software was to start them at an earlier level, with the goal of providing some remedial support. I was able to identify some of the obstacles they were encountering. These particular two students weren't as independent as my other students. I think these kids had a history of struggles with math and that they were not persisting in their learning. If they got something wrong, they would just sit there and wait for me to come around.

I also noticed they were having particular difficulty with problems that had multiple steps. The assessments that Khan Academy™ provided helped me realize what specific issues they were having. But the intervention didn't involve using Khan Academy™.

Jim: What did you do?

Jeff: I did some one-on-one work with them, and I also used another tool called Light Bot™. It's interesting because it works like a Lego™ Robotics kit. If you haven't seen how those work, students have to build a robot and program the robot to guide it's movements. They program the robot to turn left here, make this many steps forward, turn right, etc. Light Bot™ is a video game-based version of that. It helped me understand the thought process of each of the students because it wasn't a math problem, per se, but with related concepts and spatial understanding. Concepts like "What is left?" "What is right?" and counting. With these students, I found that just the ordering of steps was enough of a challenge for them. And that was beneficial. What I noticed was one student would create instructions for the video robot to take one step at a time and then would run the program. He would check to see if the robot got to the target and then write instructions for the next step. He needed to progress step-by-step and needed feedback after every step.

Jim: How did you help him?

Jeff: There was guesswork in trying to understand why he worked that way and finding ways to encourage him and also advance him to the next stage. I helped him by asking, "Why don't you try three steps at a time," and I discovered that there was a little bit of reluctance to make a mistake. I wanted to help him get over this reluctance; to be successful in school, you have to take intellectual risks. I realized that years of not being successful in school and math might have caused him to be very cautious. He improved with the three steps strategy, and when he made a mistake, I would reassure him and urge him to stick to the new strategy. I think there was progress and there was more willingness to make mistakes. I wish I had more than one year with him because that was all I saw by the end of the year. It was a great learning experience for me and something I can share with other teachers. I think it helped the student gain a little bit more confidence. Even though we didn't get him back on grade level by the end of the year, I think he was willing to engage a little bit more.

Jim: How about the other student?

Jeff: The other student exhibited, at first, what seemed like similar issues. The results were similar; they would get the same grades on a test. That also highlights why an assessment tool like Khan Academy™ can be so helpful, because when I saw a more personalized profile of that student, I realized how different they were. That student was able to do recite multiplication tables fluently. But, there was a focus issue. If the student was completing a test or even completing an activity on Khan Academy, particularly if it was a word problem, he might read halfway through the problem, then lose focus and make up an answer. On a test, there would be sections left undone. I used a similar intervention with the Light Bot™, and for this student, I saw quick improvements. The immediate feedback he was getting from the Light Bot™ kept him engaged. What I ended up developing with him was using the technology as a reward to break up a class period of work. I think Light Bot™ was valuable for him in developing executive functions (the ability to self-regulate). I did see progress from him, and I think part of it was increased engagement. At the beginning of the year, he didn't feel like he had a chance of success in the class.

Jim: What do you think the turning point was?

Jeff: There was the focus issue, and then there was a motivation problem. They were related, and they were limiting him. I think a big part of it was just getting him to buy in. Getting him to trust me as a teacher, the fact that I was reaching out and giving him other options. And, I think it did build his executive functions, his focus, and his ability to do things in sequence. I don't know if he quite demonstrated grade-level math ability by the end of the year, but he did a better job of showing what he already knew how to do. Using Khan Academy™ helped because I was able to identify a starting point for these students. The software got me to a point where I could start trying out different things. In previous years, I may not have had enough time to dedicate to one individual student, and also I may not have been able to identify that one student was working on second-grade material and the other was working on third-grade material.

The interactive, Internet-based Khan Academy™ software provides a variety of video-based learning experiences and activities that assess what students are learning. The software provides students with multiple attempts at problems and provides targeted information based on what types of challenges and activities the student succeeds at and those that present a struggle. The "real time" nature of the software allows the teacher to target the interventions in the specific ways that Lisciandrello describes. This type of strategy offers many advantages over paper and pencil assessment strategies, including the fact that many students view the interactions with the computer as fun. Lisciandrello provides the Light Bot™ software as fun and interactive mathematical activity, and this provides opportunities for meaningful interactions and math discussions about programming the Light Bot™.

Jim: How many kids do you have in class when the interventions were going on?
Jeff: 16 – 18 students.

Jim: And this is in the private school right?
Jeff: Private school, upper class.

Jim: When you were working with the students independently, what were the other children in the classroom doing?
Jeff: Mostly what we would do on Khan Academy™ days was group work in table teams. There was a group doing advanced algebra, another group doing moderately advanced work, and there was an "on grade level" team. There were also kids who were trying to catch up to grade level, the remedial group. The two struggling students were not able to keep up with the remedial group.

Jim: Khan Academy™ makes it harder for you to incorporate the different kids into a group that is performing at a well-defined level. It makes it clear to students that they can't keep up, which keeps them out of the social group environment. How do you handle that?
Jeff: That's true for those group work days, and that's why I don't use these tools for 100% of class time. What we would do is alternate. I would do two days a week of personalized learning and two days a week of project-based learning. If you used a method that did not highlight the differences among students, the students who had a greater need would not get that support. I've seen group projects

where "all" the students are participating in the group project, but actually one or two students are carrying the whole team. There's always that balance of how much do you want to call out student differences, and how much do you want to hide them for the students' comfort. I think there is a space for a small amount of discomfort.

Jim: How did you group students?

Jeff: When we did project-based learning (PBL) activities, I think that gave more opportunities for students at all levels to interact with each other. One activity was about creating a programming project that moved a chess piece on a digital game board. Some of the students who were very successful in traditional assessments of math performance would struggle with that project. Sometimes, the students who struggled with the traditional assessment of math performance did well with programming. Creating visual models would be another example of a group-based project. Some students could zip right through long division and multi-step word problems, but if I asked them to draw a visual model of five divided by two, they became lost. A student who didn't seem like a strong math student became very successful with visual models. The culture I try to create in the classroom is that it's ok not to be able to do something. We're going to do enough different things so that everybody is going to have a chance to be the expert.

In his descriptions of the students' responses to his interventions, Lisciandrello notes that they responded differently to his attempts at interventions. In the first case, Lisciandrello realized technology was not enough to support the student, and his active participation was required. In the second case, the technology created an opportunity for student engagement in an activity that produced, according to Lisciandrello, self-motivation. Lisciandrello's uses of technology enabled him to look at the situation of his students in new ways and created new possibilities and "new wants" in the classroom. Lisciandrello wanted to invest time in individualized support because the technology produced the opportunity be more efficient at using assessment data. The students responded to Lisciandrello's attention and that created new wants for the students. Lisciandrello uses terms like *remedial* and *advanced group* as part of normal teacher discourse at school. Grouping arrangements do reveal differences among students. Students can only get individualized support when teachers can demonstrate that there are differences that are significant enough to justify additional assistance. The issues with grouping students seem to be less evident when students are allowed to work collaboratively in project-based learning activities where different approaches to thinking and expression are encouraged.

Many of my graduate students talk to me about using technology to personalize learning and differentiate instruction. Carole Anne Tomlinson is often credited as being a leader in the differentiated instruction movement. Tomlinson is an American educator who has written extensively about techniques in differentiated instruction. Many professional development workshops are built around her books and materials. However, attempts to make differentiated instruction a widespread approach to teaching date back to the 1890s.

During the 1890s, it was difficult for teachers to address the needs of individual children and, as a result, they were socially promoted, which created difficulties in classrooms of the time (Washburne 1953). The problem was that schools grouped children according to their chronological age, when, in fact, educators realized that children learned and developed at different rates within a chronological age grouping. According to Washburne, in 1890 Preston Search, a Pueblo Colorado educator, addressed the problem by making "self-instruction" possible in school. It was a radical idea that was well ahead of its time. Washburne's historical account reveals that the self-instruction movement eventually gave rise to self-instruction materials in textbooks and workbooks in the early 1900s. The idea behind self-instruction was to make schooling fit the child, not the other way around. The child would progress at his or her own pace. This revolutionary moment in education gave way to the established institutional order of the time. Another major innovation of the 1890s was standardized testing, an innovation that also created the concept of the "average student." Washburne documented other education reforms that still impact today's' school settings. Ability grouping, mental maturity grouping, interest grouping, all failed to produce the desired goal of creating a method for getting children in the same chronological age group to move along at the same speed. In his 1953 paper, Washburne concluded that the solution to the differentiation dilemma was to create a common core curriculum and to identify the achievement outliers. The outliers would have opportunities for self-learning (in isolation) until reintegration was possible. The majority of teachers would spend most of their time working on moving average students through a core curriculum.

The idea that school should fit the child and the child should progress at his or her pace is appealing. We know from Vygotsky's studies that children in the same chronological age group may be at different levels of cognitive development and may have different zones of proximal development given the same assistance (Vygotsky et al. 1978). Jeff Lisciandrello discovered this when working with the two boys and technology in his math class. Differentiation attempts to develop different entry points to the same concept or the same learning task. For example, a visual presentation is substituted for a textual one when the task is to identify the characteristics of the geometric figure. The goal of differentiation and most classroom instruction is to get everyone to be able to perform the same task or know the same facts. Differentiation is an old idea that keeps getting recycled and is now being paired with instructional technologies to personalize learning. In contrast, to paraphrase Holzman, we can create learning communities with many tasks and no goals (Holzman 2008).

That's Why Math Learning Is Hard

In this chapter, we have only scratched the surface of the various challenges in math learning. The performative approach prioritizes building relationships to math as a social activity, which includes conversations, project-based learning, and performing math in everyday contexts. Mathematics learning is a social activity that includes "math talk" and working with technology to produce mathematical artifacts like the Math Video Project that would transform mathematics learning in classrooms.

Many of my graduate students learning about my Math Video Project will say things like, "It sounds like such a fun and exciting thing to do, but I don't have the time." I remind them that the developmental approaches to teaching and learning that I propose are best accomplished with the support of others, including students in the classroom. Teaching math in a way that transforms it into a social activity should be both collaborative and celebratory.

References

Askew, M. (2008). Maths with meaning. Retrieved from http://mikeaskew.net/page3/page2/files/Mathswithmeaning.pdf.

Askew, M. (2011). Private talk, public conversation. King's College London. Retrieved from http://www.mikeaskew.net/page3/page5/files/Privatetalkpublicconverse.pdf%5Cnhttps://www.learntogether.org.uk/Resources/Documents/PrivateTalkpublicconversationMikeAskew.pdf.

Askew, M. (2016). *Transforming primary mathematics: Understanding classroom tasks, tools and talk*. New York: Routledge.

Askew, M., Brown, M., Rhodes, V., & Johnson, D. (1997). *Effective teachers of numeracy* (pp. 0–126). London: King's College. Retrieved from http://www.mikeasew.net/page4/files/EffectiveTeachersofNumeracy.pdf.

Cole, M., & Consortium, T. D. L. (2006). *The fifth dimension: An after-school program built on diversity*. New York: Russell Sage Foundation.

Gallistel, C. R., & Gelman, R. (2005). Mathematical cognition. *The Cambridge handbook of thinking and reasoning* (pp. 559–588). Retrieved from http://doi.org/10.1037/002775.

Holzman, L. (1997). *Schools for growth: Radical alternatives to current educational models*. Mahwah, NJ: Lawrence Erlbaum Associates.

Holzman, L. (2008). Creating stages for development: A learning community with many tasks and no goal. In A. Sumaras, A. Freese, C. Kosnick, & C. Beck (Eds.), *Learning communities in practice*. NY: Springer.

Lave, J. (1988). *Cognition in practice*. New York: Cambridge University Press.

No Child Left Behind (NCLB) Act of 2001, Pub. L. No. 107–110, § 115, Stat. 1425 (2002).

Rogoff, B. (1990). *Apprenticeship in thinking: Cognitive development in social context*. United States: Oxford University Press.

U.S. Department of Education. (2013). The Nation's Report Card: Trends in Academic Progress 2012. NCES 2013–456. *National Center for Education Statistics*. Washington D.C.

Vygotsky, L., Cole, M., John-Steiner, V., Scribner, S., & Soberman, E. (1978). In M. Cole, V. John-Steiner, S. Scribner, & E. Soberman (Eds.), *Mind in society: The development of higher psychological processes*. Cambridge, MA: Harvard University Press.

Washburne, W. (1953). Adjusting the program to the child. *Educational Leadership*, December, 139–150.

Yelland, N., & Kilderry, A. (2010). Becoming numerate with information and communications technologies in the twenty-first century. *International Journal of Early Years Education*, *18*(2), 91–106. Retrieved from http://doi.org/10.1080/09669760.2010.494426.

PART III

Development Takes Practice

In the final part of this book, civic engagement and performance activism in the form of service-learning pedagogy are discussed. Chapter 8 details the interdisciplinary service-learning project that I have been leading for the last 5 years. Chapter 9 features reflections on my search for method in STEAM education.

Inspiration

I met Fernanda Liberali, a Brazilian teacher-educator and Vygotsky scholar, several years ago. She was presenting with her students and her colleagues at the Performing the World Conference (PTW) that is hosted by the All Stars Project and the East Side Institute for Group and Short Term Psychotherapy in New York City. It was an ensemble presentation where the professors and the students, who were part of a teacher preparation program, told stories of going into the community to teach English in the *favelas* (slums) in São Paulo, Brazil, during their teacher-training program. I was struck by how unified they were. They presented as a collective, and they closed their presentation with a song. During the question-and-answer portion of the session, someone asked where they got their funding. Liberali responded that there was no provision for the community-service aspect of the teacher-preparation program. The students (a dozen or so) were all volunteers and they had done the fundraising to pay for the trip to the conference. Later, at a post-conference social event, I asked Liberali how she had gotten her students to volunteer. I don't remember exactly what she said, but she left me with the following impression: It takes organizing, insistence, and creativity to get things done in bridging university work to community work. Liberali was an inspiration to me. At the time, I had been working as a public school teacher in the South Bronx in New York City, and I had a sense of connection to the community work she and her students described. Liberali used Vygotsky and was integrating Newman and Holzman's ideas about performance and community activism, and she gave me a sense of new possibilities in education.

Service-Learning

Service-learning is a type of experiential learning. It takes students out of the classroom and provides the opportunity to apply concepts learned in the classroom to support the needs of a community (see Chap. 2). Service-learning partnerships with public schools is not a new idea, and many universities have well-developed, large-scale service-learning programs that do many wonderful things for communities. Many kinds of service-learning projects send college students into public school classrooms. These projects tend to be oriented toward college students with an interest in becoming teachers. That is a worthy goal in service-learning, and there is a strong possibility that some of the students on the projects we created may be inspired to teach. Our interest in service-learning is to create developmental learning environments in education institutions, not as a reform, but as a radically different approach to learning that is both inclusive of traditional forms of learning and transformative.

Activity That Creates Public Good

In Chap. 1, a review of Michael Teitelbaum's book on STEM workforce development helped to demystify whether or not America is falling behind in the global race to employ scientific talent. Of all of the very useful things that Teitelbaum has to say on the subject of STEM education and science, perhaps the most useful is the reminder that science contributes to the public good. The public benefits from scientific contributions, so it makes sense for the government to use public funds to support scientific research and education research that contribute to the good of all. There are many examples of science research that improves public health and well-being: cancer, Alzheimer's disease, infectious diseases, environmental cleanup, alternative energy, and gene mapping are just a few of the areas of science research that benefit many of us. The political discourse on education in the USA is, at its foundation, about competitiveness and economic dominance; this is the education for workforce development paradigm (see Part I, Introduction and Chap. 1). When Teitelbaum refers to "public good," the sense of the term is economic. The economic sense of the phrase *public good* is attributed to Paul Samuelson from his 1954 paper, "The pure theory of public expenditure," in which he describes "collective consumption goods." The theory attempts to create a model that can be used to make decisions about spending government tax revenues on goods and services that benefit the public. Put simply, "collective consumption goods" benefit everyone. Consumption of the good does not result in less of the good being available. It is in this sense that service-learning is a social activity, supported by educational institutions, that enables students to contribute to the public good—and that is a "public good" in and of itself.

Breaching Barriers to Participation

Even though the popular press and specialized scientific journals make the dialogue public, government leaders, business leaders, and the scientific community are having conversations about STEM and STEAM learning among themselves. I don't believe that there is a conscious intent to exclude the public; this is simply an observation of how the current arrangements, leaders, and experts talking to each other may unintentionally create exclusion. Parents, students, and teachers have little direct input into policy dialogues. When there are opportunities for input, we cannot assume that a meaningful dialogue has occurred. For example, there are many occasions for town hall-style public forums and opportunities to respond to surveys that count for collecting input from the public. However, surveys limit how people can reply. Modern town hall meetings are tightly controlled, informal political interactions that are not designed to result in decisions; they are merely for information and expression of opinion and are not politically binding. Finally, as noted in Part I, STEM education and workforce readiness is a dialogue among experts who use specialized language and concepts to communicate. Informed and meaningful participation means that we are all working to communicate and create meaning together. In my opinion, we do not need institutional permission to create a new understanding of learning and STEAM education. We need to support the leadership of educators, students, and parents who want to contribute to STEAM education.

Organizing STEAM Learning Around the Public Good

If we return to the idea of science and education as contributing to the public good, we may find a way toward creating an inclusive and meaningful dialogue about STEAM education. Most people have an understanding of the philosophical sense of "public good" as something that is shared at a societal level, like clean air and clean water. Access to a free public education is considered a "public good." According to the United Nations Convention on the Rights of the Child, children have a right to a mandatory and free public education. Most governments around the world consider a free education to be a global public good.

A "public good" does not describe an infinite resource. In twenty-first-century America, we are well aware that natural resources are limited and must be protected. A "public good" implies that ethical decisions about resources are being made. The concept of *public good*, whether economic or philosophical, is useful. The creation of "public good" involves social activity, collective choices, and whether it is economic or political, it means that people are coming to a consensus on something. I believe that organizing STEAM education around the public good does some important things.

First, it reorganizes the conversation about STEAM learning in a way that is inclusive of education interests that are not aligned with economic or national goals without excluding them. Second, everyone can participate in conversations about the "public good" that they are benefiting from. Alternatively, people can also participate in discussions about how they lack access to the benefits of "public goods," such as being educated in a collaborative, developmental interdisciplinary STEAM learning environment. Public good, at the grassroots level, allows students, parents, and teachers to participate in deciding what the STEAM education priorities are and how we can go about teaching and learning. Reorganizing the STEAM education conversation as being in the interest of the public good gets us past the policy debates and questions about why are students are not ready for competition in the global workforce and on to questions concerning how STEAM education can benefit everyone.

Performance Activism

Performance activism is an idea that comes out of the joint development of an international community of educators, researchers, therapists, artists, and community organizers who have studied Newman and Holzman's Vygotsky or who are leaders in the performative turn in social activism (Freidman and Holzman 2016). Performance artists, educators, professionals, and researchers come together in New York City every two years for the Performing the World Conference (PTW) hosted by the East Side Institute and the All Stars Project. I have been an attendee and presenter at this conference since 2003, and I've met many educators from all over the world, many who use performance and Newman and Holzman's social therapeutic approaches. It is a conference where everyone shares performance-based and interdisciplinary work they are doing in communities around the world.

There are hundreds of attendees, but the atmosphere is more like a block party than an academic conference. Most of the conference sessions are held in theater spaces and meeting rooms at the All Stars Project's headquarters on 11th Avenue and 42nd Street in New York City. There are also sessions that take place around New York City at school auditoriums and Social Therapy centers. PTW is an international community of practitioners, and yet, it is highly personal and relevant. I enjoy attending and presenting at the PTW conference.

CHAPTER 8

Service-Learning Project

Organizing in Academia

I started my current appointment as an assistant professor in 2011. I was an untenured faculty member, in my late forties, on a tenure track, teaching teachers to use technology in the classroom. The experience of working in academia was different from the public schools and corporate environments that I had worked in for most of my professional life. I had to create a new performance, and fortunately, I had plenty of life experience to draw on. My scholarship and teaching were grounded in Vygotskian cultural performatory approaches to learning, but there were no other Vygotskians to be found on the faculty at my school. Recognizing this, I made maintaining and developing relationships outside of my school a priority. It was a decision that has served me well.

Yes, and

My method for navigating through my first year in academia was to introduce everyone I met to my "Yes, and" performance. "Yes, and" is an improvisation exercise that I play to teach people how to create a collective story. In faculty meetings, "No, but" or "Yes, but" were typical responses to new ideas and suggestions. I made it a point to offer "Yes, and" as an alternative response when my colleagues interacted with me. I did this in a playful way, explaining that "Yes, and" was a straightforward method for building collaborative environments. Like many new

J.E. Martinez, *The Search for Method in STEAM Education*,
Palgrave Studies In Play, Performance, Learning, and Development,
DOI 10.1007/978-3-319-55822-6_8

faculty members, I found academia intimidating, and I tended to say "yes" to suggestions to volunteer for committees or take on additional work. The amazing thing about "Yes, and" is that it creates new possibilities.

Improv Games

I accepted an assignment in my first year as a faculty member to create a professional development program for an elementary school in Harlem. The school had U.S. Department of Education Magnet Schools Assistance program funding. The funding could be used to pay for teacher professional development. This school had contacted my institution, and I was sent to work with them. I created a program that introduced many of the same instructional technologies that I was using in the graduate program that I taught in. I also included improv games that I used to create new performances and ways of being in the classroom. My friend and mentor Carrie Lobman had written *Unscripted Learning: Using Improv Activities Across the K-8 Curriculum* (2009), and I used many of the activities in the book in the professional development work. Leading strangers in improv games is not as easy as it looks, and I was encouraged that no one rejected my improvisational offers. Formal teaching observations and feedback from workshop participants were also positive and encouraging.

The professional development program with the Magnet school had me committed to one Saturday morning a month at the school for 10 months and a weeklong summer institute. Ellen Darensbourg was the school's Magnet specialist and my primary point of contact. I worked with Darensbourg to plan the workshops, train the teachers, and debrief the sessions afterward. After several months of working together, we developed a trusting working relationship that would become the foundation of our subsequent efforts in STEM (Science, Technology, Engineering, Math) education and service-learning.

Building an Environment with Relationships

"Yes, and" and other improv games are methods for connecting with people, leveling the playing field, and creating an environment for new possibilities to emerge. Everyone is uncomfortable at the beginning of an improv game, and an individual's academic rank or area of expertise does not provide a competitive advantage the way it might in a faculty or business meeting. Having a group experience that provides everyone with an equal opportunity to contribute to the group's efforts is necessary preparation for developing the capacity to collaborate. The initial work with the Magnet school was considered successful by my institution; it brought non-tuition revenues to the university. I eventually managed to expand my efforts to include additional Magnet schools. Darensbourg and I worked on continuing to expand our efforts together, but I was also bothered by my experience with teacher professional development. I did not see much development going on despite our best efforts. Teachers learned to use technology and would demonstrate that they learned it, but they were not changing their classroom practices, and their attitudes toward technology use in the classroom did not change. To them, the technology was an add-on not an essential part of the classroom experience. They insisted on having to know what to do with the technology before trying to use it. They were not comfortable playing around with the technology or exploring it. They did not have the same views toward technology that I knew children had. Children learn technology developmentally in the same way that they learn language; they do not have to know things about technology before using technology. A few hours a month in professional development wasn't producing much developmental learning. The teachers insisted that they were not comfortable allowing students to use technology in the classroom that they did not understand. From my perspective the teachers were unintentionally getting in the way of developmental learning.

Creating New Stages

Sometimes you have to risk walking on to an empty stage when you are invited up from the audience. Despite my efforts to create new practices for teachers in using technology, they were merely learning how to use technology and sometimes applying the knowledge. One day, Darensbourg explained to me that she believed that the goal of STEM education was to teach children to think like engineers, scientists, technologists, and mathematicians. I didn't see how that would be possible given my experience with teaching teachers to use technology. I thought like a technologist, yet teachers do not think about technology or use technology the way I do. I was concerned that I'd failed to create a zone of proximal development (ZPD) where thinking like a technologist or learning developmentally was possible (Vygotsky 1978). The idea of teaching children to think like a scientist was appealing, even though I didn't believe it was possible to teach children to think like a scientist or technologist. As I saw it, a developmental approach to learning should create opportunities to engage in activities that might be developmental precursors to thinking like STEM professionals do. Children could pretend to be scientists and engineers in the same way they pretend to be Mommy or the Teacher. They needed to interact with STEM professionals in STEM activities to learn to create those performances, just like they did at home.

I decided that I liked the challenge. I wanted to do something to create a developmental learning environment, a ZPD for interdisciplinary STEM education. Could we create the environment where learning to think like a scientist or engineer was possible in school? I knew that my journey to STEM included some curiosity about technology, formal training in computer science at college, extensive use of computers on my own time, and immersion in a professional technology culture. I think like a technologist (and like a Vygotskian), and there was no way that a schooling experience could reproduce a process that had developed over 20 years or to create a significant shortcut.

I had learned from my work with the All Stars Project that youth development is produced on a literal new stage. The new stage could be an All Stars Talent Show Network stage in a public school auditorium somewhere or in a corporate boardroom down on Wall Street. I knew that I needed to create stages and organize audiences to come to the show. That stage had to be a place where elementary school children could participate in shared activities with people with STEM knowledge and practices. I just didn't know how to get STEM people like scientists and engineers into an elementary school in a consistent and meaningful way. I don't recall how much time I spent mulling the idea over (it may have been a couple of months), but the answer came unexpectedly in the early spring semester of 2012.

In February, Fran, the Director of the Center for Teaching and Learning (CTL) at NYIT invited me to a student demonstration of a course capstone project in the School of Engineering and Computing Sciences. I was impressed by how the young engineering and computer science students presented their project. They were talking about Gantt charts and the challenges of planning their project, and the technical aspects of building technology-based solutions to problems in a collaborative environment. They sounded like the technologists and engineers I had worked with in my former career. I asked the tough questions, and I was satisfied with the answers. At that moment, a new idea occurred to me.

At the end of the presentation, I congratulated the students and asked Fran and the course instructor if it was possible to have engineering students show up at a public school to work with children. The instructor could not help me, but Fran suggested that I meet with Amy Bravo. Bravo is the Director International Education and Experiential Learning at the university; it turned out that her office was down the hall from mine, but we had not yet met. I scheduled a meeting with Bravo, and I gave her a copy of my previous book as an introduction and preparation for our meeting.

It was an unusual meeting for me because it seemed that 5 minutes into explaining what I wanted to do she stopped me and told me that the way to place students who were studying engineering and computer science at an elementary school was through service-learning. I asked, "What's that?" and the rest, as they say, is history. In Bravo, I found fellow community organizer and an instant friend, by the time we had completed our first meeting, we were already finishing each other's sentences and making plans to find a course for me to co-teach as a service-learning course in the School of Engineering and Computing Sciences. I'd discovered my stage (a service-learning course), my performers (college students, teachers, and children), and my producer (Bravo). Now we had to organize everyone to get to the show.

Producing School

Between March 2012 and September 2012, Bravo, Darensbourg, and I figured out how to bring students in the School of Engineering and Computing Sciences into elementary schools. We embedded service-learning into a freshman Career Discovery course. When I imagined bringing undergraduates into the elementary school, I had been thinking of juniors and seniors, but we were dependent on what there was to work with, not what we thought was ideal. There was no grant money associated with this project; Bravo and I asked academic deans to contribute from their discretionary budgets to cover the travel expenses of the students between the campus and the school. Bravo negotiated all of the administrative details with the dean of the School of Engineering and Computing Sciences. Darensbourg worked with her administration to convince them that it was a good idea for 25 college students to visit an elementary school for 10 weeks. The college students would be traveling in small teams and pairs on different days of the week and scheduled to work for an hour on each visit. I convinced my academic dean that service-learning was going to be part of my research portfolio and that I would publish the work.

The idea was that the undergraduates would arrive at the school at regularly scheduled times and participate in activities. The freshmen could choose to take part in one of three ways: the classroom hands-on learning project teams in grades pre-K through 5, fixing and upgrading equipment on the tech team, or video recording and photographing the experience with the documentary team. The work that the college students did in the school was in addition to the course work that they were expected to complete. I taught the career discovery aspect of the course during one of the two weekly sessions with the college students with my co-instructor who taught the academic engineering content.

During class time, I created a performance that was familiar to me and one that might be useful to the college students. I had been a corporate technology project manager in corporate America, and I thought that skill set would be valuable to put on display. I became the project manager of all the different projects that the college students were working on in the school. Darensbourg, the teachers, and the children were the clients. The college students comprised the project teams assigned to different aspects of a large-scale integration project. Each team had a project leader and various responsibilities assigned to each member, and each team had different tasks, and we were all working toward creating a public performance.

We had a large group meeting at the beginning of the semester before the students started service-learning in the schools. The visit began with the teams traveling to the school and doing a neighborhood walk to get familiar the route to the school. Darensbourg was introduced and provided an overview and history of the Magnet schools movement. Bravo asked the students to consider the civic engagement aspect of our work and asked them questions about what they observed about their new surroundings. She asked questions about creating change and creating connections to the community.

The college students signed up for project assignments and scheduled the days they would visit the schools. Darensbourg, Bravo, and I had agreed at the beginning of the semester to share our work at a community showcase event. The plan was to invite the entire elementary school—100 children, their teachers, and parents—to the event. Every week when I met with the teams in the class, we would discuss project tasks, milestones, and progress toward being ready for the Showcase Event. I had project leaders reporting on progress and presenting problems with projects and challenges with getting teams to collaborate. The students felt the pressure of the new demands on them. I gave performance directions and started to see new performances of collaboration, communication, and creating projects.

Darensbourg worked to track 25 college students coming and going to the elementary school every week. Service-learning students visited the school on three different days of the week, some teams in the morning hours and some in the afternoon. It was chaotic, but there seemed to be enough that was positive to keep everyone engaged in the project. Bravo provided general support to the project and kept encouraging my teaching and organizing efforts during our debriefing sessions together to discuss weekly progress. We were changing everything about what it meant to learn in formal schooling settings, and it was stressful and hard to determine whether anyone was learning from week to week. Everyone was out of their comfort zones, and we (Bravo, Darensbourg, and I) had nothing but trust in each other. I also had a Vygotskian theory of human development that suggested that putting young adults in classrooms with children could produce positive things (development) if they were engaged in meaningful activity with each other.

The tech team fixed computer equipment, upgraded software, and helped teachers put it to good use. Members of that team felt that it was wasteful to have so many computers in disrepair, and they were enthusiastic about getting them into service in the classrooms. The teachers figured out that they could create more group activities and use more technology with college students in the room. That provided many opportunities for the children to work in small groups with technology together with the college students. Teachers noticed that the kids were more excited to be in school on the days the college students visited. The engineering majors started making suggestions about engaging engineering activities, and the documentary group was capturing the enthusiasm of the school community as the changes were occurring. We would eventually discover that the changes were tangible and visible to outside observers.

On December 10, 2012, we held our Showcase Event. Four college students from our class took to the auditorium stage with two teachers, and I moderated our performance of a panel discussion for 100 children, teachers, parents, faculty, deans, and the provost of the university. The college students had produced a 15-min-long documentary video, and we had a forty-minute panel discussion that was covered by an education reporter for WNYC Radio in New York City (Fertig 2012). It was the first time that the teachers and the students had ever been on a panel on stage in front of an audience. It was the first time that the elementary school students had ever been in a college auditorium. We had produced 10 weeks of school, performing as project teams and learning about being civically engaged in a public school, and everyone was happy with the results.

Zones of Proximal Development in STEM Learning Environments

"This work is messy." That's what Bravo says when she describes our efforts in creating these new kinds of learning environments with a civic engagement learning component. The messiness is not limited to the chaos of a large project with moving parts. Many emotions are also experienced. Emotional development happens in groups that struggle together at many different levels (Holzman 2009, pp. 26–37). Working with people when they are struggling is "messy" when compared to the highly scripted approaches to schooling that is common when the primary task is to achieve specific learning goals. I find working with the messiness rewarding, and I grow personally and professionally.

We (our service-learning community) experienced many different emotions from the beginning to the end of our project. What became apparent through the documentary video and the discussion on the panel was that although we did not understand the exact impact on the children academically, we had achieved high levels of enthusiasm for being at school and high engagement in all kinds of learning activities. One surprising result, at least to the teachers, included the children looking forward to being in school and doing project-based learning with the college students. The teachers had not expected to be able to utilize the undergraduates so well. They discovered that having an "extra pair of hands" in the classroom was useful. The college students assisted with small-group learning, providing individual attention to children and using more technology in the classroom. Learning activities varied from investigating the stability of structures by building with blocks, using Lego Robotics™, and learning to use computer-aided design software that rendered 3D images of objects.

The college students reported that they had a new respect for teachers and how hard they worked. They enjoyed working with the children and connecting with them on many different levels. Several college students identified with the kids on a personal level and stated that the children reminded them of themselves at the same age. Many college students performed more service hours than were required. Some had even been able to find jobs as interns at the school after the course ended. There were many different outcomes from this work. It all happened in a messy and creative process.

We had discovered that everyone experienced lots of uncertainty at the beginning of the service-learning project. Week by week, plans that had been laid out early in the semester unraveled because of changing conditions at the school and in the lives of college students. We came to value the uncertainty that the project produced. People who run public schools do not typically appreciate the kinds of change that disrupt normal routines. Administrators expect the lessons to happen at a particular time of day. Students are required to internalize expectations of an orderly day and are supposed to know what the academic and behavioral expectations for them are. Schools operate on the underlying assumption that a systematic process will result in student learning.

Despite the messiness, most of our college students reported that they had grown from the experience and that the children they worked with benefited in ways they could observe. The Showcase Event, our ensemble performance, made our growth and development visible and obvious to audience members, including school district administrators. The Magnet school district officials were so impressed by the Showcase Event that they asked for a meeting. That meeting resulted in an invitation to scale up our project as part of the next round of Magnet grant funding to schools in Jamaica, Queens, New York, from 2013 to 2016.

The 10 weeks of service-learning allowed us to create a collective experience for college students, teachers, and children. Our work together was a social and emotional experience as well as a cognitive experience. Teachers were surprised at the emotional impact that the college students had on the children, and they explained how they were disappointed that the experience seemed to end too soon. The college students reflected on how they were taken aback by the demands of the project and the immediacy of the benefit of their presence. They felt proud that they were making a difference. Darensbourg, Bravo, and I were exhausted but proud and amazed that we had pulled all of it off, and that it all looked so good when presented in the video and on stage. When I first proposed to the college students that there was an opportunity to be part of a panel on stage, I didn't have many volunteers and had to convince the students who did participate that they could do it. Immediately after the showcase as I congratulated students, I could hear college students who had chosen not to take part in the panel talk about how they felt that they missed out on an opportunity.

Holzman writes that "the zpd is taken to be a dyadic relationship of assisting rather than the collective activity of creating" in most contemporary Vygotskian approaches (Holzman 2009, p. 29). I believe that we undertake the "collective activity of creating" when we do service-learning in schools. The college students were not merely mentoring or tutoring the children with scripted materials. They were creating conversations, interacting in projects, telling stories and asking the children to talk about their lives and interests. Everyone became more creative, and they felt good about it. The college students were role models and creators of learning environments where STEM learning was happening and technology was used.

Since that first service-learning project, there have been several others, and college students consistently underestimate their impact on the school environment. They will report on how enthusiastic the elementary school students are and how ready to learn and smart they are. I remind them that the enthusiasm that they have observed is what they have created with the children and that it is not typically the case that elementary school children in underserved communities are eager to be in school or considered smart and ready to learn. In this zone of emotional development (Holzman 2009), building enthusiasm for being in school creates different performances for children. Those new performances also convinced college students they were smart, ready to learn, and prepare for STEM careers. *Taking pride in someone else's new performance is an emotional development for a college student and a college professor.*

Being and Becoming

I lived in a continuous state of worry while I worked on the confusions, the missed project dates, and the obstacles, and at the same time, I was confident. The college students knew that they had my support when things did not go well at the school. I was becoming confident that we were creating an environment where development was happening. I had a feeling from the weekly reports and conversations I was having with the college students that learning was happening. I was confident the Showcase Event would be transformative, and that we would see the transformation at the event. I was confident because I was creatively imitating a model of learning and development that I knew very well.

The All Stars Project is a national youth development program that has a 12-week leadership program for high school students called the Development School for Youth (DSY). In that program, adult volunteers work with inner-city high school students to create new performances that will prepare them for summer internships in corporations. The students, predominately Black, Hispanic and poor, would learn a "White middle-class professional performance" from the adults. Upon completion of the program, students would perform their "graduation" in front of an audience of the adult volunteers, parents, financial contributors, and internship sponsors. The program is based on Newman and Holzman's social therapeutics and the benefits of performance and creating opportunities for young people to perform. I was one of the first adults who volunteered in the program in the late 1990s. We discovered then that supporting the development of young people through performance was something that was developmental for us (adults) as well. Having had that experience, I was very confident that positioning college students to support the development of children would promote college student development. The college student "outcomes" included the development of communication skills, empathy for others, increased awareness of societal issues impacting education, and connecting career goals to the context of civic engagement. My creative "imitation" is Vygotsky's term for what is happening in a ZPD (1978, p. 87). My "imitation" of the DSY program and Newman and Holzman's approach to creating learning environments is what Holzman refers to as "a type of performance"—a "becoming," a way of taking "who we are" and creating something new (Holzman 2009, p. 31). It is an ongoing dialectical process that is developmental and creates development.

We took college students who are STEM majors and put them in an elementary school. We created something new, and it was valued. Did the children learn anything about Science, Technology, Engineering, and Math? Have they acquired the practices of STEM majors? According to the college students and teachers, the children are learning the things that they have to teach. The children are also trying to "be" like their college mentors. The following written reflection by a Magnet school coordinator is offered below (summarized and edited for clarity).

> Ms. H, one of the Magnet school specialists, reported that she was part of a conversation that elementary school students were have during a math learning session where they were practicing math skills. The children were talking about the college students they had met through the service-learning project. They talked about how they liked the way the undergraduate students worked with them and talked with them. They were excited to tell the college students how they were "working like machines," at their math stations. While solving their math problems, they discussed the college students' choice of study to become engineers and about how they needed to do well in math if they chose the same path. While they were working at their math stations, the students asked Ms. H. if she noticed that they were working in their groups in the same way they would if the college students had been there. Ms. H was surprised and thrilled to be part of the conversation with the children.

What Ms. H describes is the "math talk" and creative imitation of children who are in the "process of being and becoming" in relationship to the college students. That process might not be visible to the college students, but it became visible to a teacher. It is also a process that could not be measured using a standardized assessment. The standardized assessment does not capture enthusiasm for learning math; as a matter of fact, standardized assessment tools kill enthusiasm for math. The following is an anecdote provided by a public school administrator commenting on the impact of service-learning in one of the schools in Jamaica, Queens (summarized and edited for clarity).

> I will end this reflection with an experience one of our Magnet teachers shared with our evaluation team during a recent site visit to her school. She said that the college students have skills, talents, and insights that are quite different from her own. She went on to discuss one of her students who was having trouble in class, both academically and socially. She talked

about how this little boy was a loner, how he was often disengaged in the classroom, and did not participate in class or complete assignments. She said that he did not think of himself as "smart" or "talented" in any way. She recalled how the college students were working with this child on a hands-on STEM activity. The teacher was unfamiliar with the activity. She went on to say that the college students noticed that the child had done something particularly elegant and sophisticated in this STEM activity. During the wrap-up at the end of the lesson, the college students had complimented the boy and shared his work with the class, pointing out his skill and talent to the other children and her. The college students could not have known the impact that this recognition had on that child, but it was profound. The teacher said—and this is what I find most moving—that she would never have recognized this child's ability on her own because she was not very "tech-savvy" and only had a basic, layperson's understanding of the STEM activity. She was thankful that the college students had been in the classroom to recognize the little boy's talent and to encourage him and help him to see himself in a different way! She conveyed that this experience made her question how many other students might have talents that she is unaware of, but that might be recognized and nurtured if there were people around who had the experience and knowledge to see them. This event helped her understand in a much deeper way the importance of exposing her students to individuals and experiences outside of the classroom and the school community.

—Magnet schools project manager

Performance Activism

Bravo, Darensbourg, and I have grown our project from one school in Harlem to three elementary schools in Jamaica, Queens, New York. Darensbourg moved from Harlem to Queens with the Magnet funding and continues to be instrumental in coordinating our efforts in the schools. We've embedded service-learning into two courses in the College of Arts and Sciences that seem to be well suited to service-learning. They are titled, "Foundations of Inquiry" and "Foundations of Scientific Process," and they work to focus college students on discovering new things about learning, community engagement, and methods of inquiry in public schools. As a result of our growth and getting funding for our partnership with the schools, we have been able to run three concurrent service-learning courses in three schools with three different faculty members. I took on the role of researcher. Bravo continued to provide administration to the program and taught one of the courses. We both worked on recruiting and training other faculty to teach the courses. One faculty member, Lauren Rigney, has become so skilled that she has transformed her approach to teaching and assessment in the “Foundations of Scientific Process” course. In her class, there are no tests, and students provide weekly journals on their observations, “experiments” in the classroom, and “stories of discovery.” Rigney teaches content from astronomy, chemistry, biology, and physics and connects that content to civic engagement in STEM learning activities.

During the focus groups I conducted in each of the courses, college students take the opportunity to ask me about my interests and what I expected to happen next. I tell them that I consider public schools to be places that are anti-developmental, uncreative, and in desperate need of radical change. Service-learning provides an opportunity to reinitiate development in formal learning environments while still trying to achieve the goals of schools. What is clear to me is that what we can do in schools is learn to perform being a scientist or an engineer or a technologist or a mathematician. Children can pretend to be parents and teachers when they play among themselves, and with service-learning we have created an environment where they not merely “pretend” to be scientists, engineers, and mathematicians, but they “become” scientists, engineers, and mathematicians with young adults who are in the process of becoming scientists, engineers, and mathematicians (See in this chapter).

I hope that we can continue to grow the project and make service-learning with public schools a permanent part of the curriculum for STEAM majors at my university—and at others. We have demonstrated a cultural performatory approach to learning and creativity, and we have developed the capacity to organize communities of learners across cultures and institutions. Now we must be more insistent on getting increasing support for our efforts and broader participation from faculty. I hope that we can inspire other institutions and other educators to imitate our efforts or to join us in creating developmental learning environments. Based on feedback we have received at conference presentations, it seems we have developed a useful framework for building service-learning partnerships and creating productive and performatory environments. My work in service-learning was the foundation for my interest in interdisciplinary learning and the questions I started to ask.

Service-learning and the challenges of creating development across disciplinary, institutional, and cultural boundaries inspired my search for method in STEAM education.

References

Fertig, B. (2012). College STEM Students Offer Hands-On Mentoring. Retrieved December 5, 2016, from http://www.wnyc.org/story/303162-college-stem-students-offer-hands-on-mentoring/.

Holzman, L. (2009). *Vygotsky at work and play*. London: Routledge.

Vygotsky, L., Cole, M., John-Steiner, V., Scribner, S., & Soberman, E. (1978). In M. Cole, V. John-Steiner, S. Scribner, & E. Soberman (Eds.), *Mind in society: The development of higher psychological processes*. Cambridge, MA: Harvard University Press.

CHAPTER 9

Reflections

A Performatory Approach to Developmental Interdisciplinary STEAM Learning

Throughout the course of this book, I've presented examples of successful approaches to STEAM and STEM education initiatives. The practices and voices of educators who are interdisciplinary and innovative have been prominent. I've argued for creating alternatives to traditional educational approaches by showing that educators are improvising with and within the school systems they work in. They use traditional and non-traditional methods, and they collaborate with students and external partners. Finally, I've presented a Vygotskian cultural performatory approach to creating developmental STEAM learning environments through middle school classroom and college service-learning projects. I hope that others who are interested in STEAM education are encouraged to be performatory, playful, and creative as a result of reading this book.

Context

In my experience, interdisciplinary STEAM education experiences are collaborative, creative, and developmental ensemble performances. Like a child uttering her first words or taking her first steps, STEAM education ensembles struggle to make sense, contend with uncertainty, and seem awkward. The service-learning project featured in Chap. 8 started with uncertainty and felt awkward for a couple of years but is now an

J.E. Martinez, *The Search for Method in STEAM Education*,
Palgrave Studies In Play, Performance, Learning, and Development,
DOI 10.1007/978-3-319-55822-6_9

experience in learning that is produced by our ever-changing ensemble with great confidence. The ensemble performance of service-learning is one way that my colleagues and I create our method of interdisciplinary STEAM education. There are many other possibilities, the Cultivating Ensembles in STEM Education Research Conference, the Performing the World Conference, and the All Stars Talent Show Network presentations are other examples of bringing different people and communities together to produce cultural performatory interdisciplinary learning experiences.

The STEAM education movement presents us with opportunities to develop new attitudes about education. I framed current efforts at reform as an education for workforce development paradigm in Part I to show what limits our attitudes about education. The paradigm is the box that we are all in, and performatory approaches are outside of the box. Being out of the box is disorienting and entails a certain amount of risk taking. I believe risk taking and uncertainty are conditions that need to exist if we want to create developmental learning in institutional settings. It also helps if support, coaches, champions, and teams can be organized for being outside of the box.

It is possible to create performatory experiences that break people out of the predetermined scripts that institutions impose. STEAM education is a new mandate to introduce creativity and innovation into education. For a short time, the STEAM education movement will be open-ended, and we have an opportunity to explore and create educational alternatives and share them. Many of the existing tools we use for educational assessment were created to measure learning in traditional settings. We will need to build new tools to understand developmental learning in schools.

Successful STEAM education projects will have teams of people from different backgrounds who have all agreed to create a learning experience or project together. The roles that people take on in interdisciplinary teams will not be clear-cut since the projects they create will change as learning environments develop. STEAM educators will be the organizers, directors, leaders, and producers of developmental learning environments. STEAM educators will have to locate resources and expertise outside of schools and bring them into the school through physical and technological means. When bringing in resources is not possible, students and educators will go out into communities where the STEAM activities are happening.

Performance

Teaching is a performance; there is an audience, a loose script, and an opportunity to improvise. One of the "stage directions" that I offer to teachers is to enable student choice and participation in creating with the scripted curricula in their classrooms. Responses to that direction have varied widely, but as I demonstrate to the teachers that I teach, providing students with options makes them work harder because making choices and contributing to creating the learning environment is challenging work.

Research is a performance, and interdisciplinary research is a new kind of research performance. Researchers involved in interdisciplinary collaborations spend a fair amount of time sorting out what they are doing together. The goals may be vague at times, and the route to the goals will need to be discovered. Communication, social and emotional skills, and being open to new ideas will be critical to cultivating ensembles in STEAM education research. Fostering playful and performatory attitudes through theatrical performance and creativity exercises will be helpful to the development of interdisciplinary ensembles.

I imagine that STEAM education ensembles will build communities and will be created by different communities of practitioners coming together. Bringing together diverse groups of people who would not typically meet is, in part, made possible by twenty-first-century technologies that are now accessible to many cultures around the world. We have the technical capacity to create developmental learning environments that cross borders of every type.

I also believe that STEAM education has to happen at the grassroots community level for it to be developmental and transformative. Grassroots community building is labor intensive, socially and emotionally demanding work. Providing the leadership to create inclusive and diverse communities is an area in which we all need to develop.

STEAM educators do not work, create, play, or perform alone. That is perhaps the most significant finding that I can offer. STEAM educators and interdisciplinary practitioners are grassroots leaders who are willing to lead their students and colleagues. They succeed and fail in small ways, and they try again. They recruit peers and outsiders to their efforts, and they succeed and fail again in different ways, and their efforts eventually get noticed. All the while they are learning how to organize people and build developmental learning environments. I've observed that STEAM educators are opportunistic; they see opportunities in breaks in the institutional routines, in unique events, indifferent supervisors, and the interested ones. STEAM educators are restless innovators and are unaccepting of the way things are.

STEAM educators and interdisciplinary practitioners work on reorganizing the fundamental structures of institutions or build entirely new ones. In Chap. 4, students working with teachers transformed the use of technology in schools by contributing their knowledge and skills and created development for themselves and others. Students working with teachers created a new structure within the institution: student-led teams that fixed technology and mentored other students.

I do not believe that the knowledge acquisition approach to learning in math, science, and the other subjects will deliver the hoped for transformations of STEAM education. I don't see how introducing innovation and creativity into the education for workforce development paradigm will result in developmental learning. Why would we expect that putting more reforms into the educational reform box would result in anything other than small incremental change? The search for method in STEAM education explores a new performatory and transformative approach to education. I am an educational outsider and insider, and I'm encouraging a performatory approach to education that is interdisciplinary and comes from outside of formal education institutions. It is a methodological approach that has a history of success in afterschool youth development programs, community development projects, therapeutic environments, and formal educational settings. A performatory approach is needed for the transformations that we would all like to see and that some of us feel are necessary to bring about educational change. The teacher–student collaborative projects and descriptions of activities and approaches to learning described in this book are early indications that educators are discovering performatory alternatives and that creativity and innovation are possible even within the constraints of traditional learning environments.

Play

When people hear me talk about performance and play in schools, they look at me as if I had broken a rule. On a couple of occasions, students, colleagues, and administrators have rephrased my comments about play adding, "They can play to learn specific things." To be clear, I believe that children should be allowed to play in school (and maybe play school), and they should be encouraged to perform what they are learning. I do not believe that play needs to or even can serve specific purposes in school. If the students cannot opt out of the play activity or cannot change the activity, then whatever it is that is going on is merely masquerading as play. STEAM educators need to embrace the collaborative playfulness that is inherent in imaginary play and art. A suggestion that I can make here is one that I have benefited from: Encourage students and collaborators to participate in improvisational play, and lead the play. Afterward, have a conversation about what happened and what it felt like. I'm sure your ensemble will have discovered something new.

Imagining a TED Talk About Preparing for the Real World

There have been times in my recent career when teachers have worked with me in the classroom or read something that I wrote and then started a sentence with, "But in the real world…" I'd like to imagine a TED Talk about the real world and write a script for it. According to its Website (http://www.ted.com), TED is a nonpartisan nonprofit devoted to spreading ideas, usually in the form of short, powerful talks. TED began in 1984 as a conference where Technology, Entertainment, and Design converged and today covers almost all topics from science to business to global issues. I'd like to present one more creative performance before the curtain drops.

The TED Talks video starts with the familiar TED logo and audience applause. Martinez is standing on a darkened stage within the spotlight. In the background, a large screen displays a video clip from a service-learning project in STEAM education. Martinez raises his mic and addresses the audience.

Many people will say things like "school prepares children for real life" or "you have to go to school so you can find a job in the real world in the future." The video you just saw is only a couple of minutes long, but it is pretty clear that we brought people together who ordinarily wouldn't meet. In the video, college and elementary school students were engaged in STEAM learning activities and conversations. The video didn't show preparation for a job in the real world. What you saw was the real-life practice of valuable real-world skills. When I watch the video, I see teachers, children, and community partners together, developing and learning the skills they need to transform what it means to be in school.

I think that people relate to school as not being part of the real world because there are so many things about school that don't happen outside of school. For example, school is the only place that requires children to sit for nearly 6 h a day. No parent should wish this on a child, no matter how attractive that might sound. Also, school is the only place where adults are isolated from other adults for nearly 6 h a day. Again, parents who are at home with the kids know what that's like and often look forward to being with other adults at the end of the day. The bottom line is school life doesn't feel like the life we live outside of school. Many people might think that this is the way it's supposed to be. I don't.

I think that there is an opportunity in the STEAM education movement to recreate schooling so that we can't tell the difference between school and any other part of life in the real world. We have to create a new performance of schooling. Unfortunately, the people who run schools can't do this by themselves. They need our support to create a new performance of schooling. I like to call it a cultural performatory approach to developmental interdisciplinary STEAM learning. It's a mouthful, and I don't have a clever acronym, but we can call it a performatory approach or developmental STEAM learning.

Creating developmental learning environments is something I've been doing in afterschool and formal educational settings for the last 20 years. In these developmental environments, I've experienced how young people and students develop when they are invited to perform as leaders and learners. When I started doing this work in afterschool settings, I discovered that I was learning, developing, and performing while I was supporting young people to perform and develop.

I view the service-learning work that I do as bringing afterschool development to the school day. Our service-learning college students aren't teachers, but there they are creating learning activities with children. The college students report that they learn important things about communication, civic engagement, learning, and building relationships from interactions with children. Our teachers realize that with the extra adults in the classroom they can take on new performances as facilitators, directors, and producers of the learning environment. Everyone is supported by someone else to do something they don't already know how to do. We create developmental STEAM learning with college students volunteering for an hour a week for just 10-weeks. For a child that attends school for 6 hours a day, 180 days a year, that's less than 1% of their time in school. We know that developmental learning has an impact because we've been doing it in afterschool programs for decades. Now college students and teachers are reporting that children are benefiting from the tiny opportunities for development we are creating during the school day in our STEAM education projects. It's a win for college students, a win for teachers, and a win for children. Imagine what students and teachers might be able to do if we increased the number of hours of developmental learning with community partners in schools to 5% or even 10%.

Service-learning and afterschool programs aren't the only ways to create developmental STEAM learning environments. Teachers who have become invested in bringing STEAM education to schools have discovered that their students are great collaborators when they get to contribute as leaders and experts at school. I've seen how teachers have organized students who wanted to volunteer to become members of technical support teams and peer mentoring groups. These teachers have been creative, playful, and performatory in building relationships that create interdisciplinary STEAM education. They've been creative in spotting opportunities to change the way things are. They've been playful in acknowledging that they need help from others and that it doesn't mean that they have failed at something. They are performatory when they perform as leaders and support leadership in others. The STEAM education movement has created an opportunity for us to build new stages for new performances at school.

What does it take to get up on a stage and perform? Performing for an audience can be a very scary proposition. It takes preparation, practice, and a willingness to be playful in front of an audience. Those are the same things any good teacher can do. I encourage you to go out and find your STEAM ensemble or team and create your performance of interdisciplinary STEAM development. The good news is that there are many other educators, artists, and performers who can help. I've even written a book that can help you get started.

Some might challenge your efforts at playing, performing, or making a fun STEAM project in a formal educational setting. They might ask how playing and performing will translate into preparing students to compete for jobs in the real world. Here's what you might say: Students who develop the ability to collaborate, to see old problems from new perspectives, and to lead others in efforts to solve complex interdisciplinary problems won't need to compete. They are busy transforming the world right now.

Martinez drops the mic and walks off the stage, the TED video ends.

Curtain.

Bibliography

Chaiklin, S. (2003). The zone of proximal development in Vygotsky's analysis of learning and instruction. In A. Kozulin, B. Gindis, V. S. Ageyev, & S. M. Miller (Eds.), *Vygotsky's educational theory in cultural context* (pp. 39–64). New York: Cambridge University Press.

Civic Impulse. (2016). H.R. 2272—110th Congress: America COMPETES Act. Retrieved from https://www.govtrack.us/congress/bills/110/hr2272.

Cole, M. (1996). *Cultural psychology, a once and future discipline*. Cambridge, MA: Harvard University Press.

Cultivating Ensembles in STEM Education and Research. (n.d.). Retrieved from http://improvscience.org/cestemer.

Dossey, J. A., Halvorsen, K. T., & McCrone, S. S. (2012). Mathematics Education in the United States 2012. The National Council of Teachers of Mathematics, Inc.

Editorial Projects in Education Research Center. (2011, July 7). Issues A–Z: Achievement gap. *Education Week*. Retrieved from http://www.edweek.org/ew/issues/achievement-gap/.

Gordon, E. W., Bowman, C. B., & Mejia, B. X. (2003). *Changing the script for youth development: An evaluation of the all stars talent show network and the Joseph A. Forgione development school for youth*. Institute for Urban and Minority Education, Teachers College, Columbia University.

Holzman, L. (2013). Are you a performance activist? Retrieved from http://loisholzman.org/2013/07/are-you-a-performance-activist/.

Lave, J., & Wenger, E. (1991). *Situated learning: Legitimate peripheral participation*. New York: Cambridge University Press.

Lowell, B. L. L., & Salzman, H. H. (2007). Into the eye of the storm: Assessing the evidence on science and engineering education, quality, and workforce

J.E. Martinez, *The Search for Method in STEAM Education*,
Palgrave Studies In Play, Performance, Learning, and Development,
DOI 10.1007/978-3-319-55822-6

demand. *The Urban Institute*, (October), ii, 48. Retrieved from http://doi.org/10.1037/e723632011-001.

Martinez, J. E. (2011). *A performatory approach to teaching, learning and technology*. Rotterdam, The Netherlands: Sense.

Newman, F., & Holzman, L. (2006). *Unscientific psychology: A cultural-performatory approach to understanding human life*. New York: East Side Institute.

New York City Dept. of Education Office of Curriculum, Instruction & Professional Learning. (2016). *STEM Education Framework*. New York. Retrieved from http://schools.nyc.gov/NR/rdonlyres/DE2FC1DE-5FB8-474F-BD27-D75FF70EF610/0/STEMframework_WEB1.pdf.

Office of the President of the United States. (2014). *Progress report on coordinating federal science, technology, engineering, and mathematics (STEM) education*.

Rogoff, B. (1990). *Apprenticeship in thinking: Cognitive development in social context*. New York: Oxford University Press.

Samuelson, P. A. (1954). The pure theory of public expenditure. *The Review of Economics and Statistics*, *36*(4), 387–389. doi:10.2307/192589.

Scribner, S., & Cole, M. (2014). Cognitive consequences of formal and informal education. *Science, New Series*, *182*(4112), 553–559. Retrieved from http://www.jstor.org/stable/1737765.

The Economist. (2010). Net generation unplugged. Retrieved from http://www.economist.com/printedition/2010-03-06.

United States Congress, U., & Coun, R. (2015). Congressional Record—House.

Vygotsky, L. (1987). In A. Kozulin (Ed.). *Thought and language* (2nd ed.). Cambridge, MA: Massachusetts Institute of Technology.

Why Interdisciplinary Research Matters. (n.a., 2015). *Nature*, *525*(7569), 305. Retrieved from http://doi.org/10.1038/525305a.

Index

J.E. Martinez, *The Search for Method in STEAM Education*,
Palgrave Studies In Play, Performance, Learning, and Development,
DOI 10.1007/978-3-319-55822-6

Printed in the United States
By Bookmasters